Student Study Art Notebook

to accompany

Foundations in
Microbiology

Fifth Edition

Kathleen Park Talaro
Pasadena City College

Boston   Burr Ridge, IL   Dubuque, IA   Madison, WI   New York   San Francisco   St. Louis
Bangkok   Bogotá   Caracas   Kuala Lumpur   Lisbon   London   Madrid   Mexico City
Milan   Montreal   New Delhi   Santiago   Seoul   Singapore   Sydney   Taipei   Toronto

The **McGraw·Hill** Companies

Student Study Art Notebook to accompany
FOUNDATIONS IN MICROBIOLOGY, FIFTH EDITION
KATHLEEN PARK TALARO

Published by McGraw-Hill Higher Education, an imprint of The McGraw-Hill Companies, Inc.,
1221 Avenue of the Americas, New York, NY 10020. Copyright © 2005 by The McGraw-Hill
Companies, Inc. All rights reserved.

 This book is printed on recycled, acid-free paper containing
RECYCLED   10% postconsumer waste.

3 4 5 6 7 8 9 0 QPD/QPD 0 9 8 7 6 5

ISBN 0-07-297808-2

www.mhhe.com

DIRECTORY OF NOTEBOOK FIGURES

TO ACCOMPANY
TALARO, FOUNDATIONS IN MICROBIOLOGY, 5/E

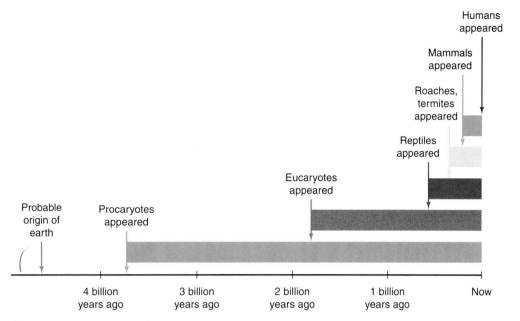

Evolutionary timeline
Figure 1.1

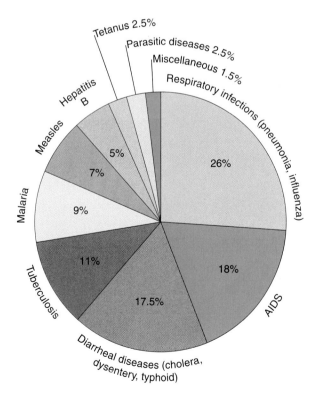

Tetanus 2.5%

Parasitic diseases 2.5%

Miscellaneous 1.5%

Respiratory infections (pneumonia, influenza)

Hepatitis B

5%

Measles

7%

Malaria

9%

26%

18%

Tuberculosis

11%

17.5%

AIDS

Diarrheal diseases (cholera, dysentery, typhoid)

Worldwide infectious disease statistics
Figure 1.4

(a) Cell Types

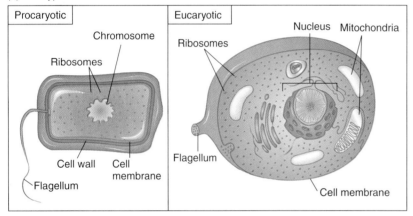

Procaryotic

Chromosome

Ribosomes

Cell wall   Cell membrane

Flagellum

Eucaryotic

Ribosomes

Nucleus   Mitochondria

Flagellum

Cell membrane

Microbial cells are of the small, relatively simple procaryotic variety (left) or the larger, more complex eucaryotic type (right).

(b) Virus Types

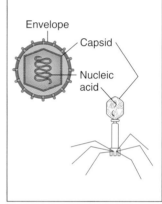

Envelope

Capsid

Nucleic acid

Viruses are tiny particles, not cells, that consist of genetic material surrounded by a protective covering. Shown here are a human virus (top) and bacterial virus (bottom).

Cell structure
Figure 1.5

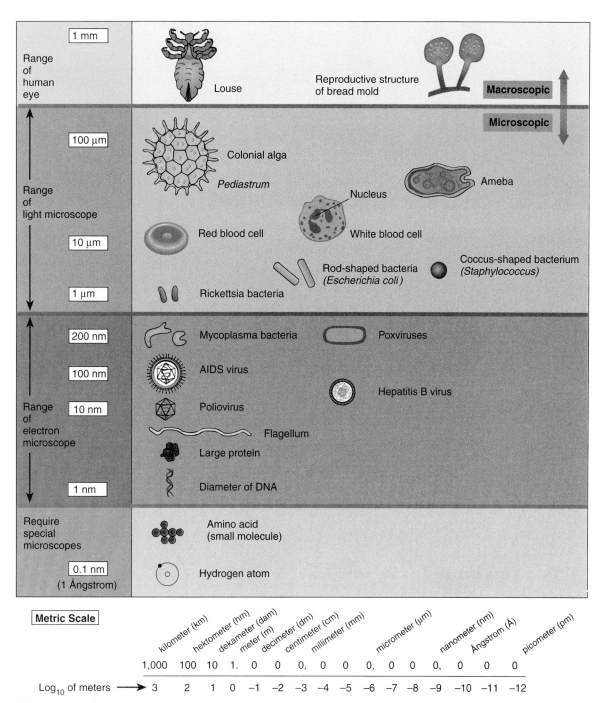

The size of things
Figure 1.7

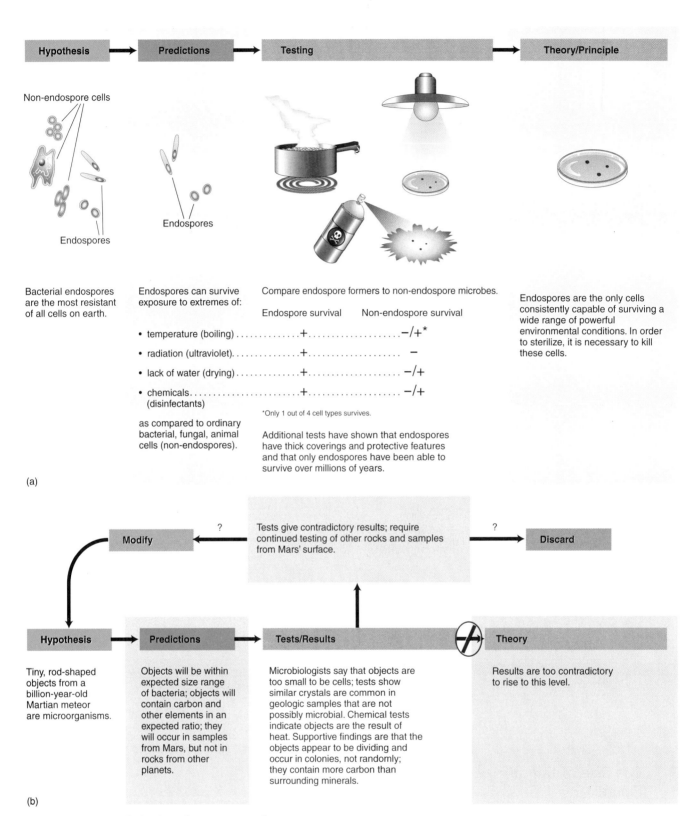

Hypothesis	→	Predictions	→	Testing	→	Theory/Principle

Non-endospore cells

Endospores

Endospores

Bacterial endospores are the most resistant of all cells on earth.

Endospores can survive exposure to extremes of:

Compare endospore formers to non-endospore microbes.

Endospore survival   Non-endospore survival

- temperature (boiling) ....+...................−/+*

- radiation (ultraviolet)..............+................... −

- lack of water (drying) ....+................... −/+

- chemicals.......................+................... −/+
(disinfectants)

*Only 1 out of 4 cell types survives.

as compared to ordinary bacterial, fungal, animal cells (non-endospores).

Additional tests have shown that endospores have thick coverings and protective features and that only endospores have been able to survive over millions of years.

Endospores are the only cells consistently capable of surviving a wide range of powerful environmental conditions. In order to sterilize, it is necessary to kill these cells.

(a)

Modify	← ?	Tests give contradictory results; require continued testing of other rocks and samples from Mars' surface.	? →	Discard

Hypothesis	→	Predictions	→	Tests/Results	≠	Theory

Tiny, rod-shaped objects from a billion-year-old Martian meteor are microorganisms.

Objects will be within expected size range of bacteria; objects will contain carbon and other elements in an expected ratio; they will occur in samples from Mars, but not in rocks from other planets.

Microbiologists say that objects are too small to be cells; tests show similar crystals are common in geologic samples that are not possibly microbial. Chemical tests indicate objects are the result of heat. Supportive findings are that the objects appear to be dividing and occur in colonies, not randomly; they contain more carbon than surrounding minerals.

Results are too contradictory to rise to this level.

(b)

The pattern of deductive reasoning
Figure 1.10

Hypothesis: Dental caries (cavities) involve dietary sugar or microbial action or both.

Variables:

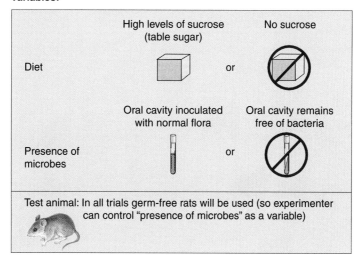

Experimental Protocol:

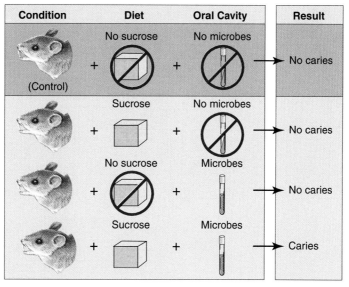

Conclusion: Dental caries will not develop unless both sucrose and microbes are present.

Variables
Figure 1.11

Domain: Eukarya (All eucaryotic organisms)

Kingdom: Animalia

Phylum: Chordata

Class: Mammalia

Order: Primates

Family: Hominoidea

Genus: Homo

Species: sapiens

(a)

Domain: Eukarya (All eucaryotic organisms)

Kingdom: Protista
(Protozoa
and algae)

Phylum: Ciliophora
(Only protozoa
with cilia)

Class: Oligohymenophorea
(Single cells with
regular rows of cilia;
rapid swimmers)

Order: Hymenostomatida
(Elongate oval cells)

Family: Parameciidae
(Cells rotate while swimming)

Genus: *Paramecium*
(Pointed, cigar shaped cells
with an oral groove)

Species: *caudatum*
(Cells pointed at one end)

(b)

Sample taxonomy
Figure 1.14

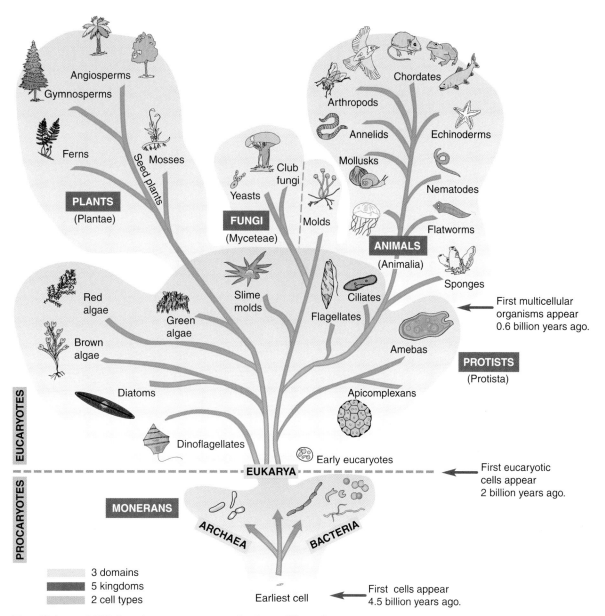

Traditional Whittaker system of classification

Figure 1.15

After Dolphin, Biology Lab Manual, *4th ed., Fig. 14.1, p. 177, McGraw-Hill Companies.*

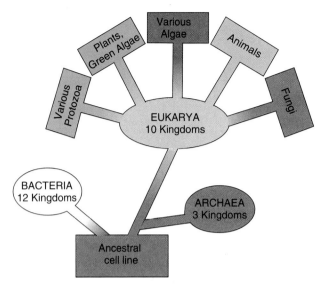

Woese system
Figure 1.16

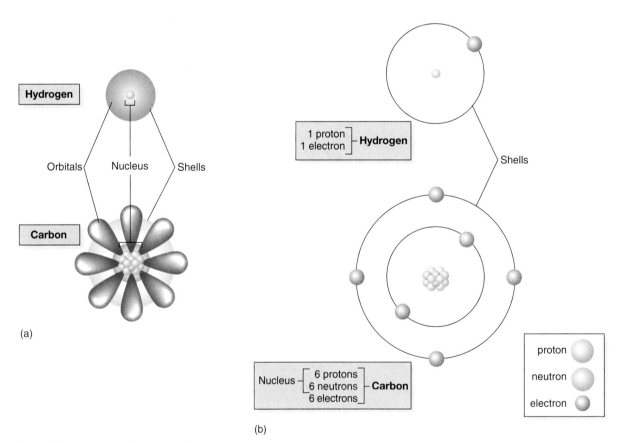

Models of atomic structure
Figure 2.1

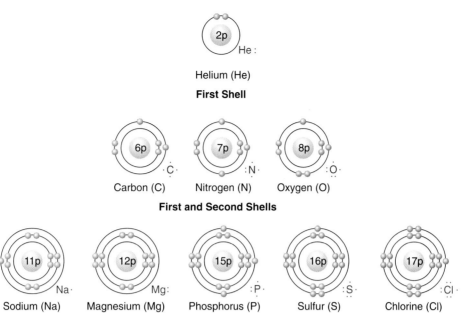

Helium (He)

First Shell

Carbon (C)   Nitrogen (N)   Oxygen (O)

First and Second Shells

Sodium (Na)   Magnesium (Mg)   Phosphorus (P)   Sulfur (S)   Chlorine (Cl)

First, Second, and Third Shells

Electron orbitals and shells
Figure 2.2

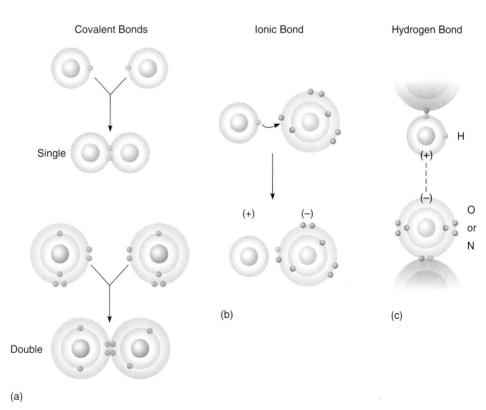

Covalent Bonds   Ionic Bond   Hydrogen Bond

Single

Double

(a)

(b)

(c)

General representation of three types of bonding
Figure 2.3

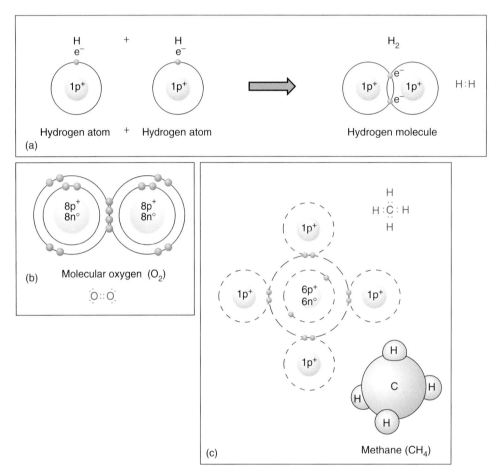

Examples of molecules with covalent bonding
Figure 2.4

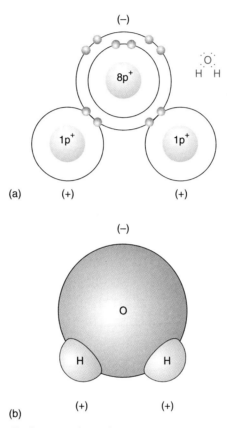

(a)

(b)

Polar molecule
Figure 2.5

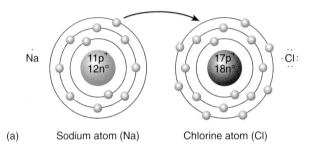

(a)   Sodium atom (Na)   Chlorine atom (Cl)

(b) Na $:\overset{\cdot\cdot}{\underset{\cdot\cdot}{Cl}}:$

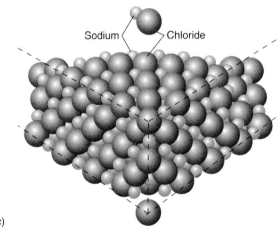

(c)

Ionic bonding between sodium and chlorine
Figure 2.6

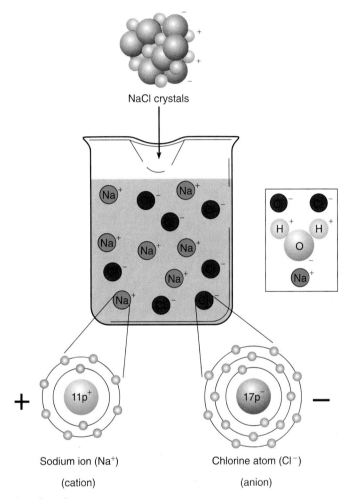

NaCl crystals

+

—

Sodium ion (Na⁺)

(cation)

Chlorine atom (Cl⁻)

(anion)

Ionization
Figure 2.7

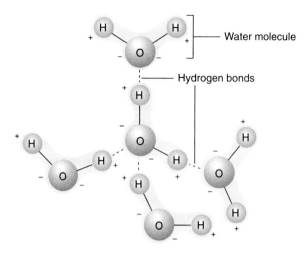

Hydrogen bonding in water
Figure 2.8

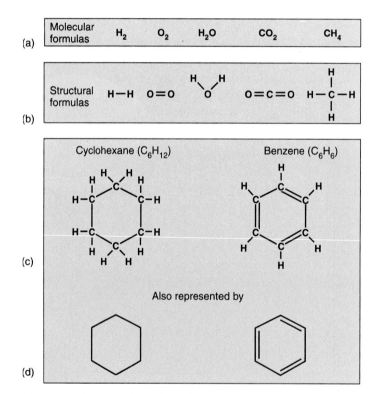

Comparison of molecular and structural formulas
Figure 2.9

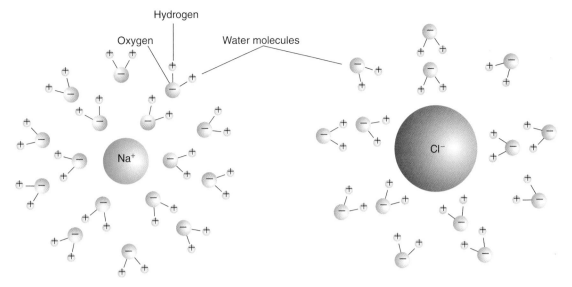

Hydration spheres formed around ions in solution
Figure 2.11

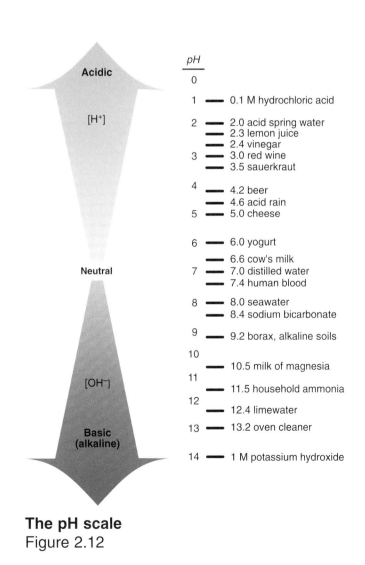

The pH scale
Figure 2.12

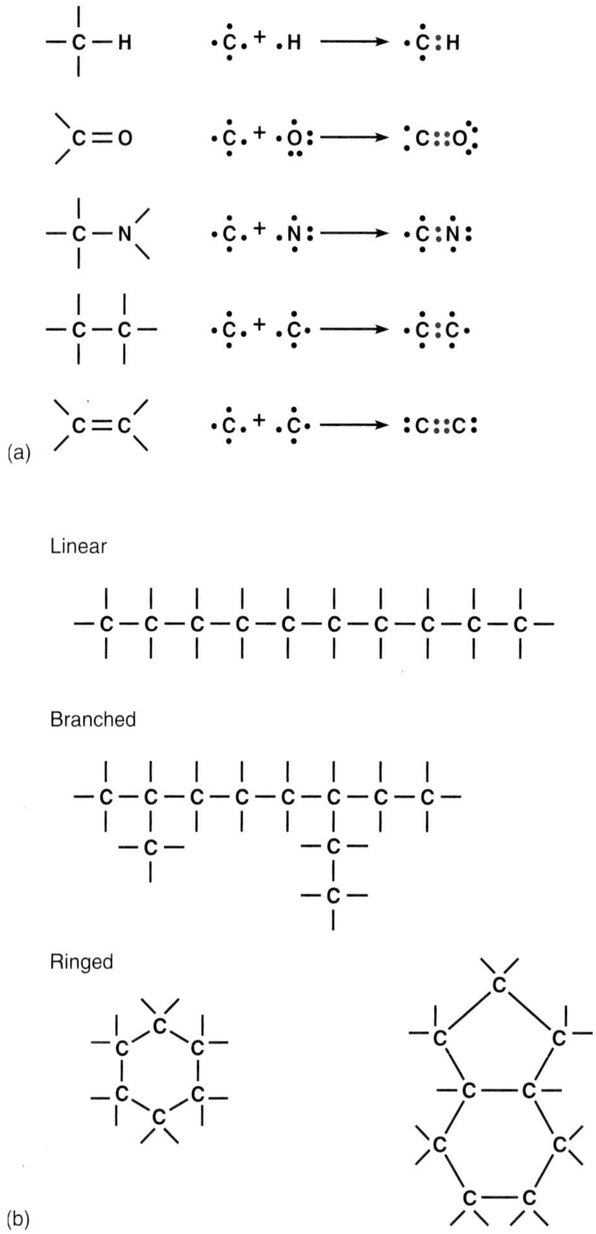

(a)

Linear

Branched

Ringed

(b)

The versatility of bonding in carbon
Figure 2.13

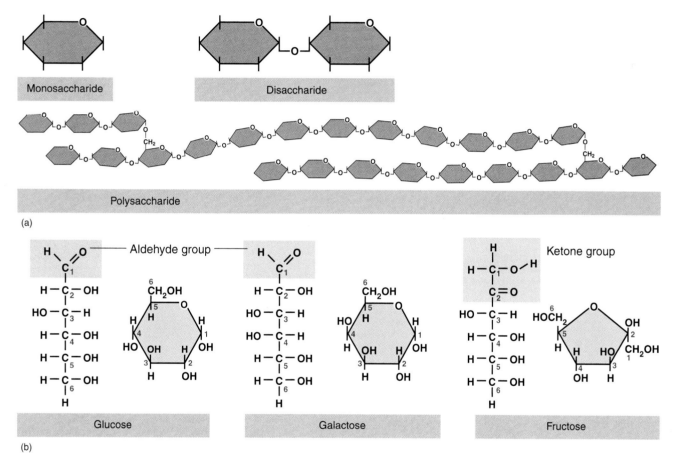

(a)

(b)

Common classes of carbohydrates
Figure 2.14

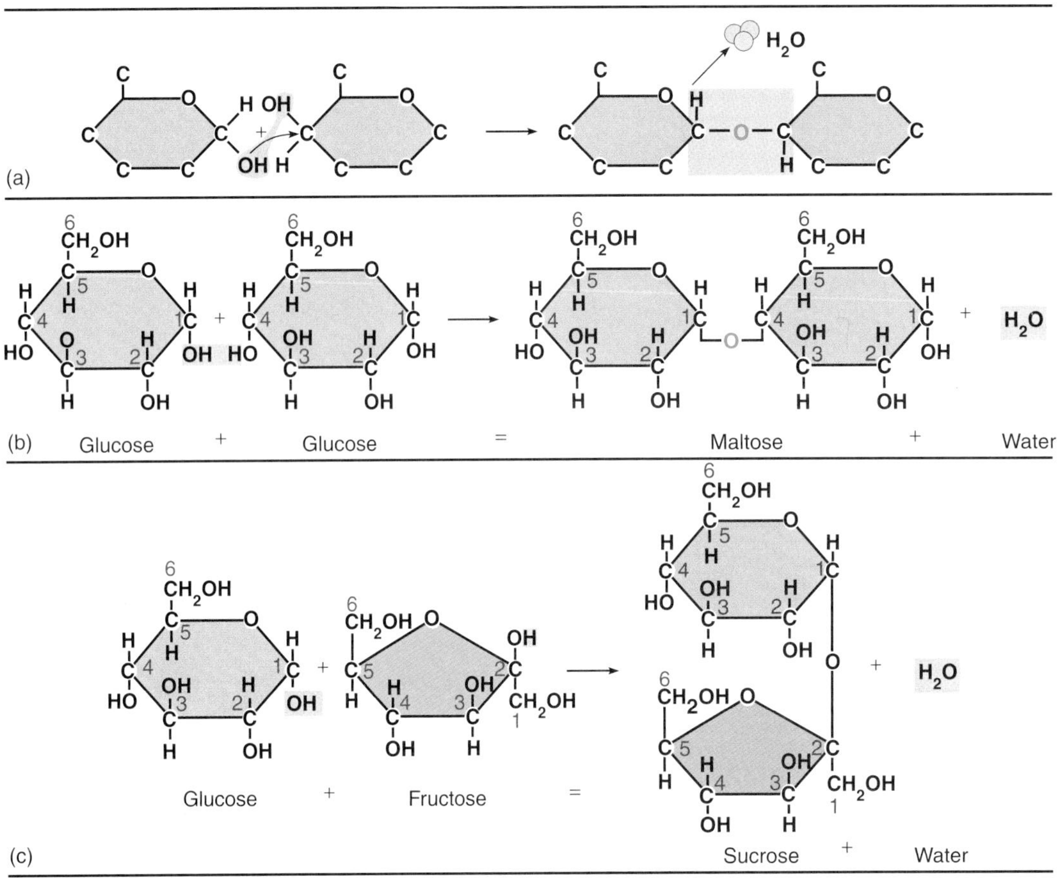

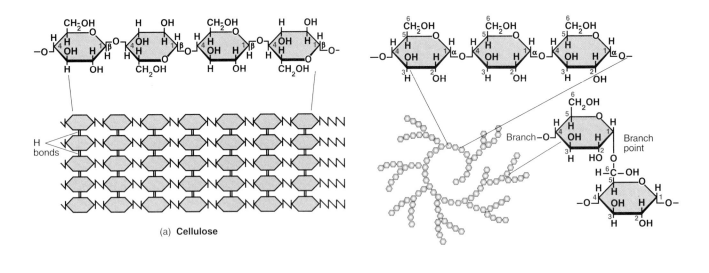

Glycosidic bond
Figure 2.15

(a) Cellulose

(b) Starch

Polysaccharides
Figure 2.16

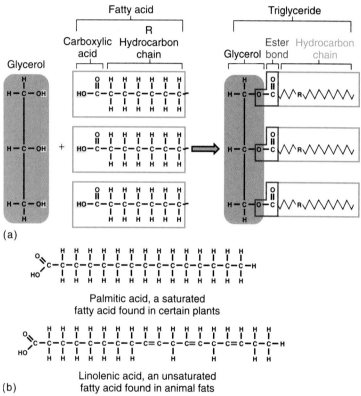

Palmitic acid, a saturated
fatty acid found in certain plants

Linolenic acid, an unsaturated
fatty acid found in animal fats

(b)

Synthesis and structure of a triglyceride
Figure 2.17

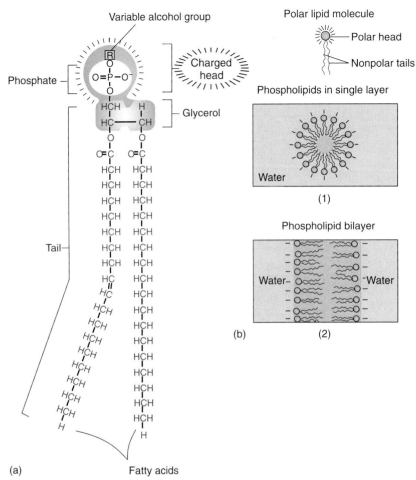

Phospholipids—membrane molecules

Figure 2.18

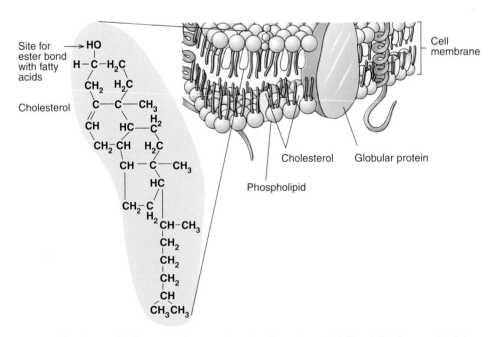

Formula for cholesterol, an alcoholic steroid that is inserted in some membranes

Figure 2.19

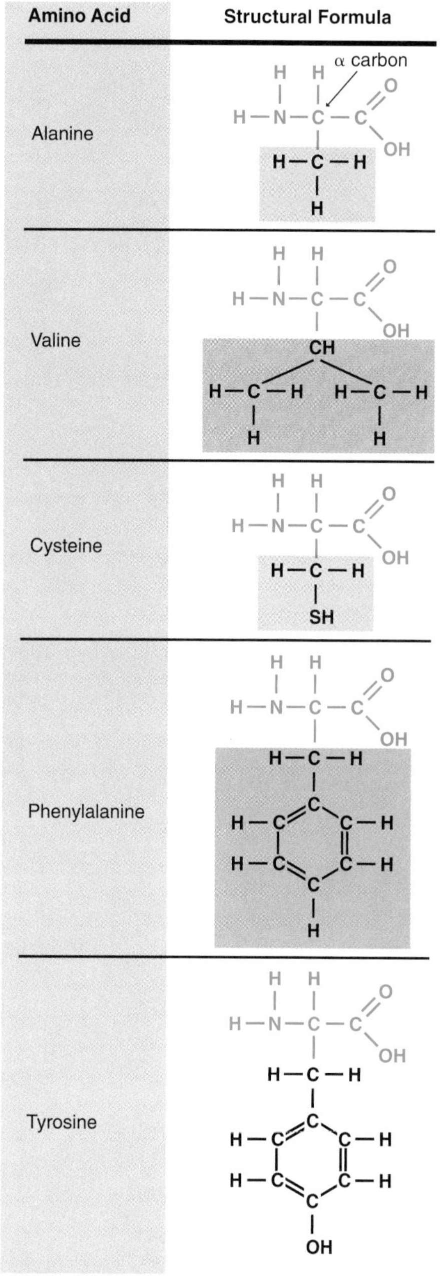

Amino Acid	Structural Formula

Alanine

α carbon

Valine

Cysteine

Phenylalanine

Tyrosine

Structural formulas of selected amino acids
Figure 2.20

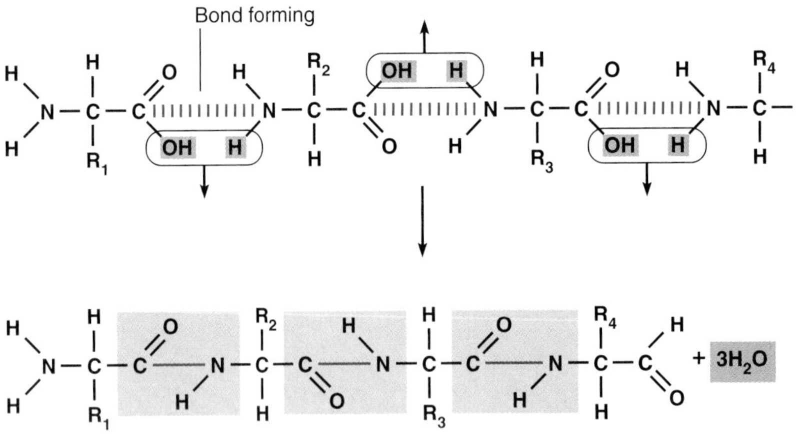

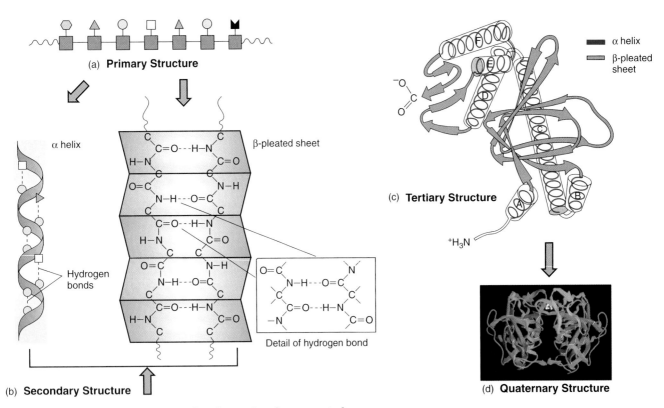

The formation of peptide bonds in a tetrapeptide
Figure 2.21

(a) **Primary Structure**

α helix

Hydrogen bonds

β-pleated sheet

Detail of hydrogen bond

(b) **Secondary Structure**

━━ α helix

▨ β-pleated sheet

(c) **Tertiary Structure**

(d) **Quaternary Structure**

Stages in the formation of a functioning protein
Figure 2.22

d: From A.S. Moffat "Nitrogenase Structure Revealed," *Science,* 250:1513, 12/14/90. © 1990 by the AAAS. Photo by M.M. Georgiadis and D.C. Rees, Caltech.

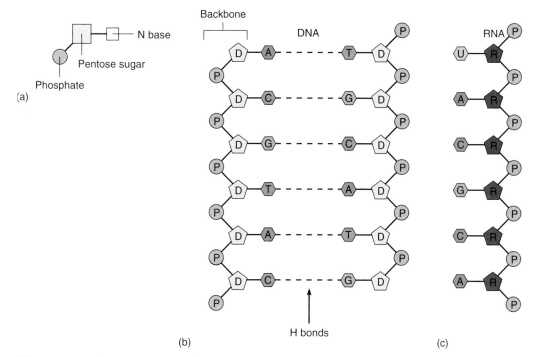

(a)

Backbone

DNA

Phosphate

Pentose sugar

N base

(b)

H bonds

RNA

(c)

The general structure of nucleic acids
Figure 2.23

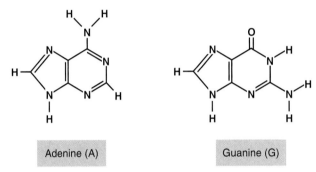

(a) **Pentose Sugars**

Deoxyribose

Ribose

Adenine (A)

Guanine (G)

(b) **Purines**

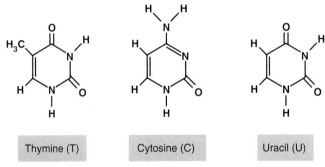

Thymine (T)

Cytosine (C)

Uracil (U)

(c) **Pyrimidines**

The sugars and nitrogen bases that make up DNA and RNA
Figure 2.24

24

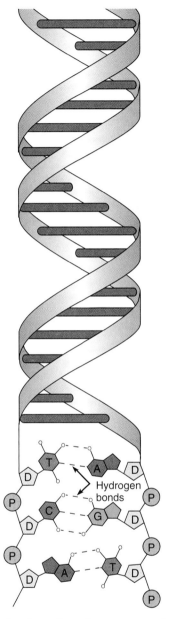

**A structural representation
of the double helix of DNA**
Figure 2.25

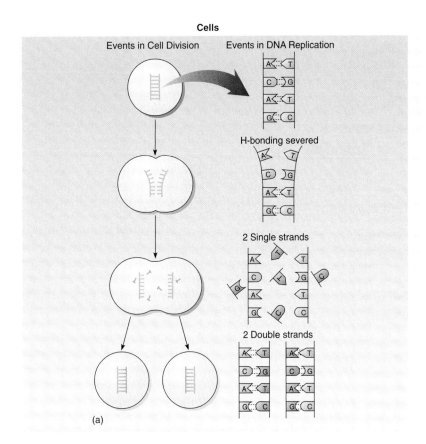

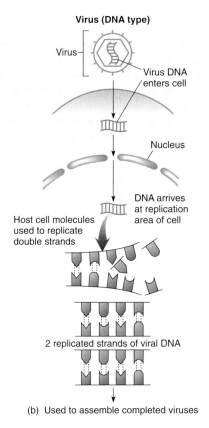

Simplified view of DNA replication in cells and viruses
Figure 2.26

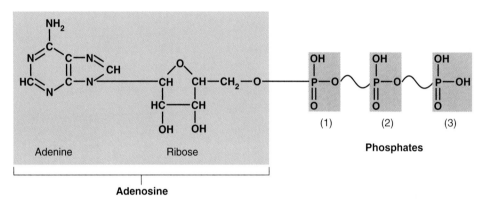

The structural formula of an ATP molecule, the chemical form of energy transfer in cells
Figure 2.27

An Overview of Major Techniques Performed by Microbiologists to Locate, Grow, Observe, and Characterize Microorganisms.

Specimen Collection:
Nearly any object or material can serve as a source of microbes. Common ones are body fluids and tissues, foods, water, or soil. Specimens are removed by some form of sampling device. This may be a swab, syringe, or a special transport system that holds, maintains, and preserves the microbes in the sample.

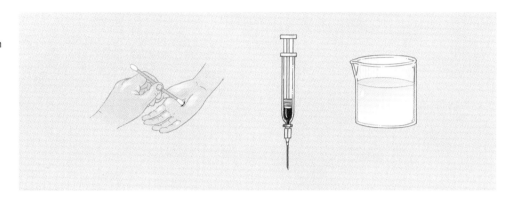

A GUIDE TO THE FIVE I'S: How the Sample Is Processed and Profiled

1. Inoculation:
The sample is placed into a container of sterile **medium** that provides microbes with the appropriate nutrients to sustain growth. Inoculation involves using a sterile tool to spread the sample on the surface of a solid medium or to introduce the sample into a flask or tube. Selection of media with specialized functions can improve later steps of isolation and identification. Some microbes may require a live organism (animal, egg) as the inoculation medium.

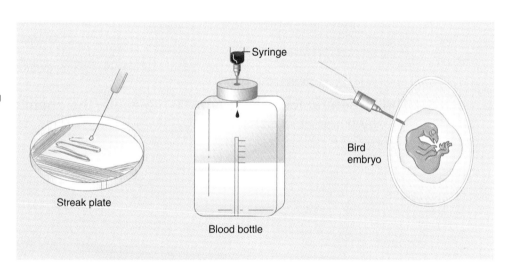

A summary of the general laboratory techniques carried out by microbiologists
Figure 3.1

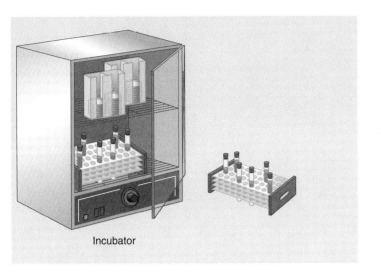

Incubator

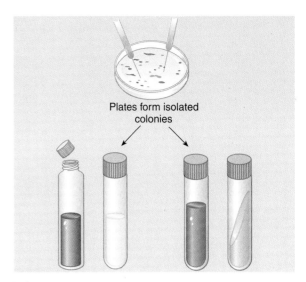

Plates form isolated colonies

2. Incubation:
An incubator can be used to adjust the proper growth conditions of a sample. Setting the optimum temperature and gas content promotes multiplication of the microbes over a period of hours, days, and even weeks. Incubation produces a culture—the visible growth of the microbe in the medium.

3. Isolation:
The end result of inoculation and incubation is **isolation** of the microbe in macroscopic form. The isolated microbes may take the form of separate colonies (discrete mounds of cells) on solid media, or turbidity in broths. Further isolation, also known as subculturing, involves taking a tiny bit of growth from an isolated colony and inoculating a separate medium. This is one way to make a pure culture that contains only a single species of microbe.

4. Inspection:
The cultures are observed macroscopically for obvious growth characteristics (color, texture, size) that could be useful in analyzing the specimen contents. Slides are made to assess microscopic details such as cell shape, size, and motility. Staining techniques may be used to gather specific information on microscopic morphology.

5. Identification:
A major purpose of the 5 I's is to determine the type of microbe, usually to the level of species. Summaries of characteristics are used to develop profiles of the microbe or microbes isolated from the sample. Information can include relevant data already taken during inspection and additional tests that further describe and differentiate the microbes. Specialized tests include biochemical tests to determine metabolic activities specific to the microbe, immunologic tests, and genetic analysis.

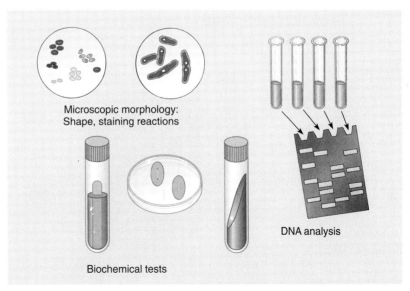

Microscopic morphology: Shape, staining reactions

Biochemical tests

DNA analysis

A summary of the general laboratory techniques carried out by microbiologists (Continued)
Figure 3.1

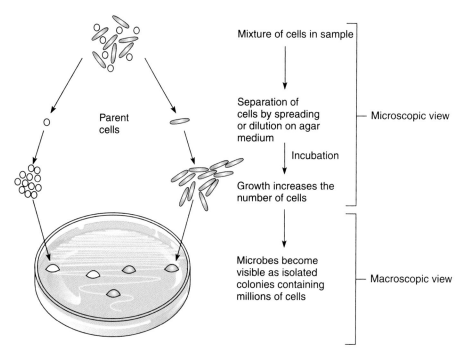

Mixture of cells in sample

Separation of
cells by spreading
or dilution on agar
medium

Incubation

Growth increases the
number of cells

Microbes become
visible as isolated
colonies containing
millions of cells

Parent
cells

Microscopic view

Macroscopic view

Isolation technique
Figure 3.2

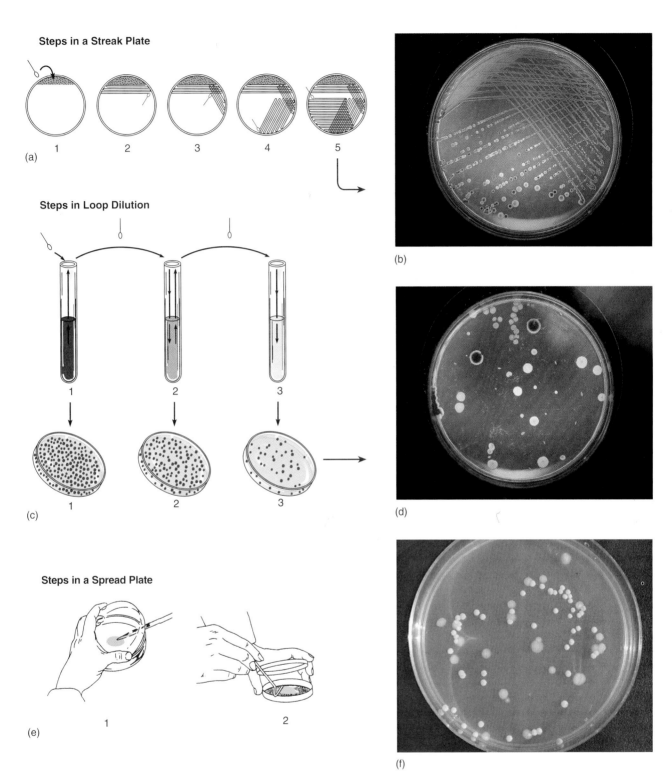

Steps in a Streak Plate

(a) 1 2 3 4 5

(b)

Steps in Loop Dilution

(c) 1 2 3

(d)

Steps in a Spread Plate

(e) 1 2

(f)

Methods for isolating bacteria
Figure 3.3
b,d,f: Kathy Park Talaro

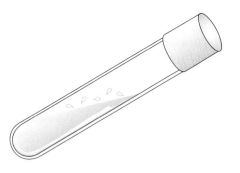

(a)

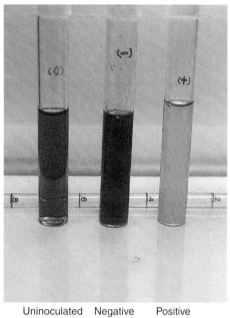

(b)   Uninoculated   Negative   Positive

Sample liquid media
Figure 3.4
b: Kathy Park Talaro

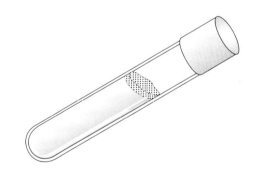

(a)

(b)

Sample semisolid media
Figure 3.5
b: Kathy Park Talaro

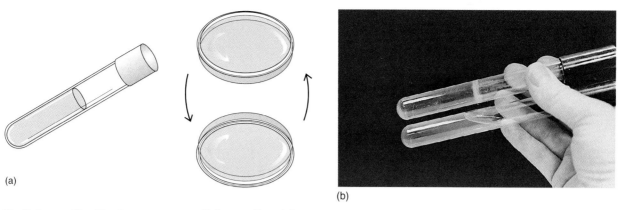

Solid media that are reversible to liquids
Figure 3.6
b: Kathy Park Talaro

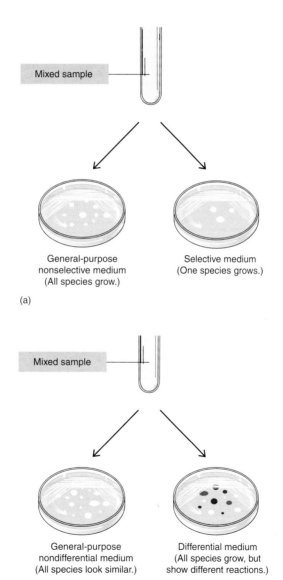

Mixed sample

General-purpose
nonselective medium
(All species grow.)

Selective medium
(One species grows.)

(a)

Mixed sample

General-purpose
nondifferential medium
(All species look similar.)

Differential medium
(All species grow, but
show different reactions.)

(b)

Comparison of selective and differential media with general-purpose media
Figure 3.8

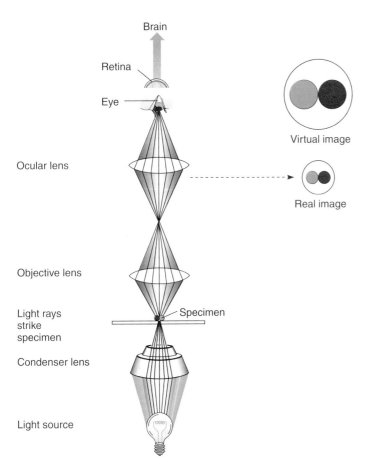

The pathway of light and the two stages in magnification of a compound microscope
Figure 3.15

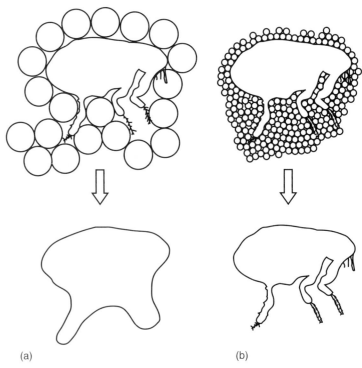

(a)   (b)

Effect of wavelength on resolution
Figure 3.16

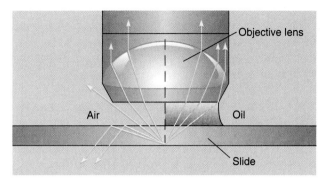

Workings of an oil immersion lens
Figure 3.17

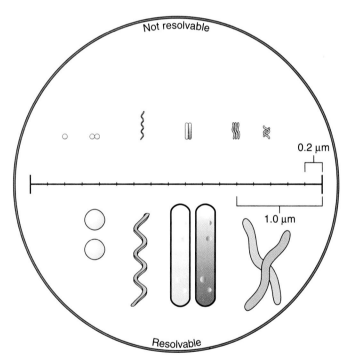

Effect of magnification
Figure 3.18

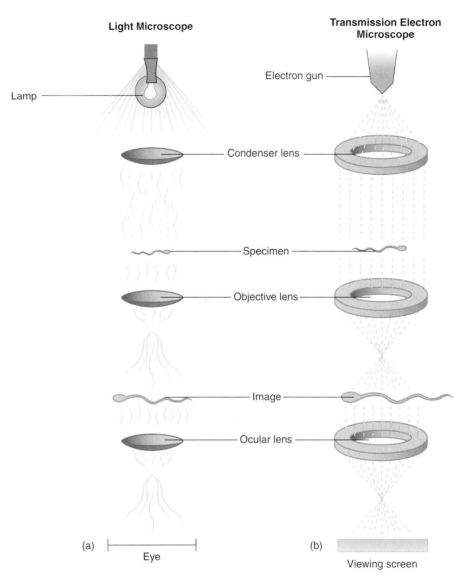

Light Microscope

Lamp

Condenser lens

Specimen

Objective lens

Image

Ocular lens

(a)

Eye

Transmission Electron Microscope

Electron gun

Viewing screen

(b)

Comparison of two microscopes
Figure 3.23

Courtesy William A. Jensen

TABLE 3.7

Comparison of Positive and Negative Stains

	Positive Staining	Negative Staining
Appearance of cell	Colored by dye	Clear and colorless
Background	Not stained (generally white)	Stained (dark gray or black)
Dyes employed	Basic dyes: Crystal violet Methylene blue Safranin Malachite green	Acidic dyes: Nigrosin India ink
Subtypes of stains	Several types: Simple stain Differential stains Gram stain Acid-fast stain Spore stain Special stains Capsule Flagella Spore Granules Nucleic acid	Few types: Capsule Spore (Dorner)

37

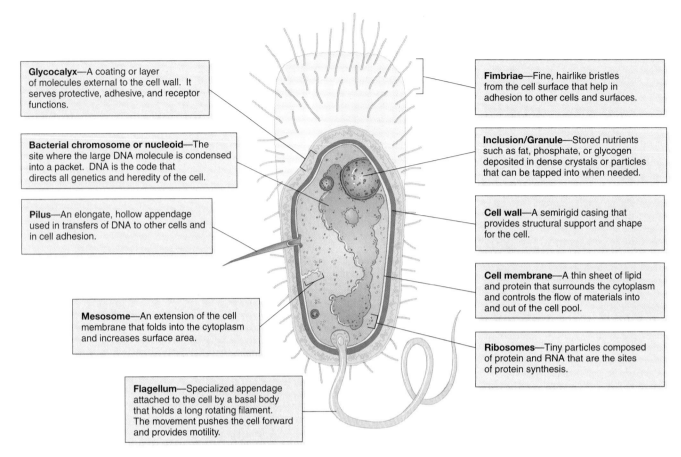

Glycocalyx—A coating or layer of molecules external to the cell wall. It serves protective, adhesive, and receptor functions.

Bacterial chromosome or nucleoid—The site where the large DNA molecule is condensed into a packet. DNA is the code that directs all genetics and heredity of the cell.

Pilus—An elongate, hollow appendage used in transfers of DNA to other cells and in cell adhesion.

Mesosome—An extension of the cell membrane that folds into the cytoplasm and increases surface area.

Flagellum—Specialized appendage attached to the cell by a basal body that holds a long rotating filament. The movement pushes the cell forward and provides motility.

Fimbriae—Fine, hairlike bristles from the cell surface that help in adhesion to other cells and surfaces.

Inclusion/Granule—Stored nutrients such as fat, phosphate, or glycogen deposited in dense crystals or particles that can be tapped into when needed.

Cell wall—A semirigid casing that provides structural support and shape for the cell.

Cell membrane—A thin sheet of lipid and protein that surrounds the cytoplasm and controls the flow of materials into and out of the cell pool.

Ribosomes—Tiny particles composed of protein and RNA that are the sites of protein synthesis.

Structure of a procaryotic cell
Figure 4.1

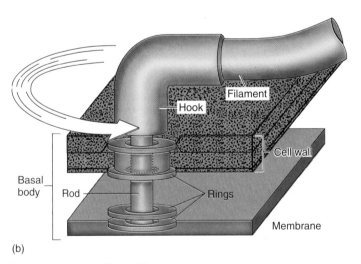

Filament

Hook

Cell wall

Basal body

Rod

Rings

Membrane

(b)

Structure of flagella
Figure 4.2

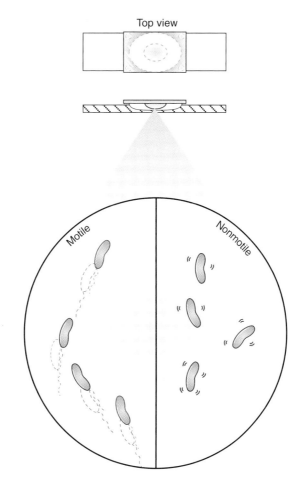

Top view

Motile

Nonmotile

Motility detection
Figure 4.4

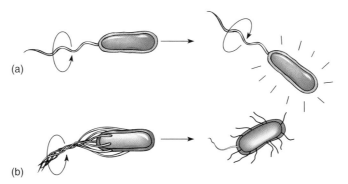

(a)

(b)

**The operation of flagella and the mode of
locomotion in bacteria with polar and
peritrichous flagella**
Figure 4.5

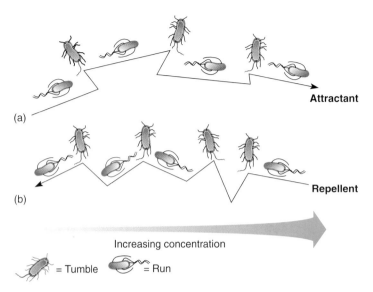

= Tumble   = Run

Chemotaxis in bacteria
Figure 4.6

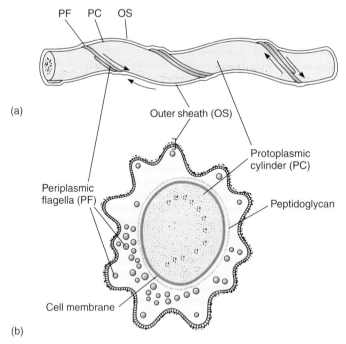

PF   PC   OS

(a)

Outer sheath (OS)

Protoplasmic
cylinder (PC)

Periplasmic
flagella (PF)

Peptidoglycan

Cell membrane

(b)

**The orientation of periplasmic flagella on
the spirochete cell**
Figure 4.7

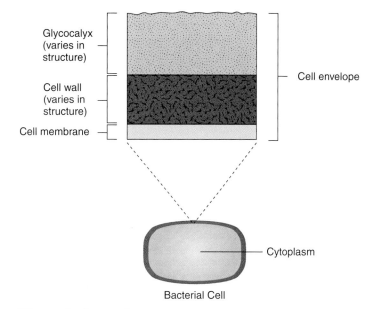

Glycocalyx (varies in structure)

Cell wall (varies in structure)

Cell membrane

Cell envelope

Cytoplasm

Bacterial Cell

The relationship of the three layers of the cell envelope

Figure 4.10

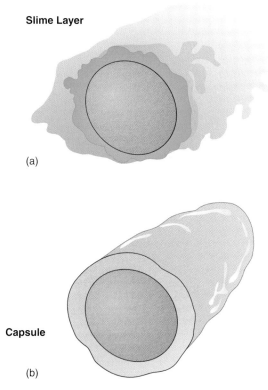

Slime Layer

(a)

Capsule

(b)

Bacterial cells sectioned to show the types of glycocalyces

Figure 4.11

(a) The peptidoglycan of a cell wall can be presented as a crisscross network pattern similar to a chain-link fence, forming a single massive molecule that molds the outer structure of the cell into a tight box.

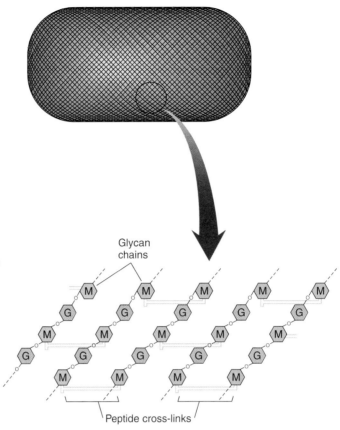

(b) An idealized view of the molecular pattern of peptidoglycan. It contains alternating glycans (G and M) bound together in long strands. The G stands for *N*-acetyl glucosamine, and the M stands for *N*-acetyl muramic acid. A muramic acid molecule binds to an adjoining muramic acid on a parallel chain by means of a cross-linkage of peptides.

Glycan chains

Peptide cross-links

(c) A detailed view of the links between the muramic acids. Tetrapeptide chains branching off the muramic acids connect by interbridges also composed of amino acids. The types of amino acids in the interbridge can vary and it may be lacking entirely (gram-negative cells). It is this linkage that provides rigid yet flexible support to the cell and that may be targeted by drugs like penicillin.

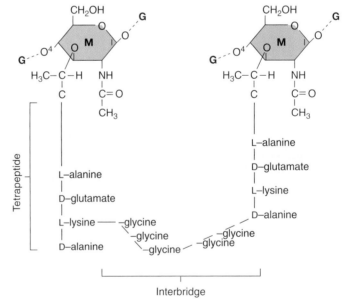

Structure of peptidoglycan in the cell wall
Figure 4.14

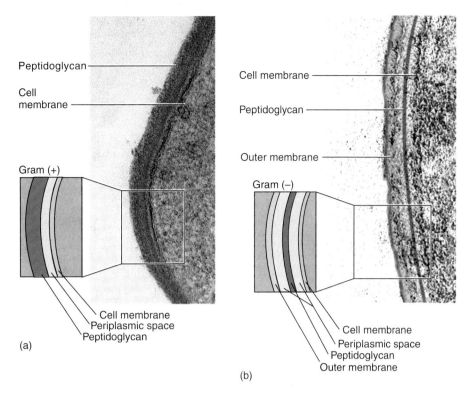

Peptidoglycan

Cell
membrane

Gram (+)

Cell membrane
Periplasmic space
Peptidoglycan

(a)

Cell membrane

Peptidoglycan

Outer membrane

Gram (−)

Cell membrane
Periplasmic space
Peptidoglycan
Outer membrane

(b)

A comparison of the envelopes of gram-positive and gram-negative cells
Figure 4.15

a2: © S.C. Holt/Biological Photo Service; **b2:** © T.J. Beveridge/Biological Photo Service

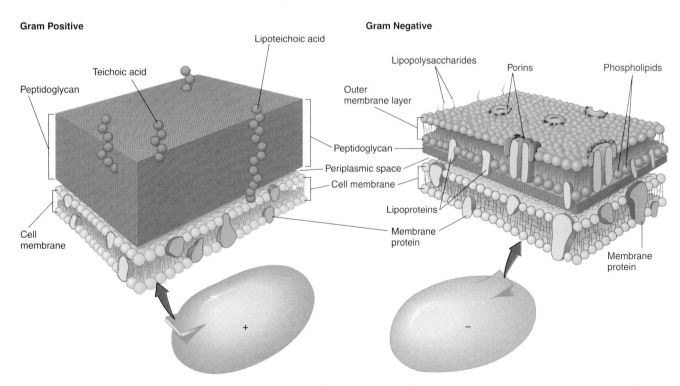

Gram Positive

Lipoteichoic acid

Teichoic acid

Peptidoglycan

Cell
membrane

Gram Negative

Lipopolysaccharides   Porins   Phospholipids

Outer
membrane layer

Peptidoglycan

Periplasmic space

Cell membrane

Lipoproteins

Membrane
protein

Membrane
protein

A comparison of the detailed structure of gram-positive and gram-negative cell walls
Figure 4.16

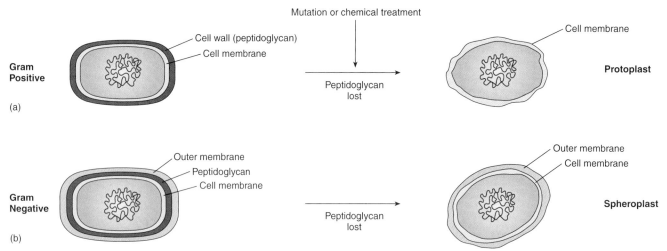

The conversion of walled bacterial cells to L forms
Figure 4.17

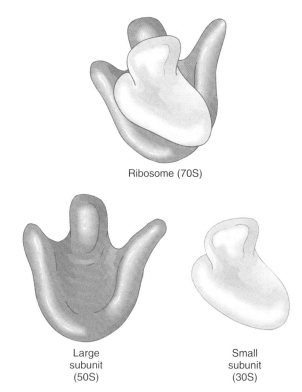

Ribosome (70S)

Large
subunit
(50S)

Small
subunit
(30S)

A model of a procaryotic ribosome, showing the small (30S) and large (50S) subunits, both separate and joined
Figure 4.19

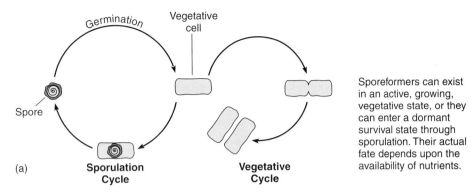

The general life cycle of a spore-forming bacterium
Figure 4.21

Sporeformers can exist in an active, growing, vegetative state, or they can enter a dormant survival state through sporulation. Their actual fate depends upon the availability of nutrients.

TABLE 4.2

General Stages in Endospore Formation

Stage	State of Cell	Process/Event
1	Vegetative cell	Cell in early stage of binary fission doubles chromosome.
2	Vegetative cell becomes **sporangium** in preparation for sporulation	One chromosome and a small bit of cytoplasm are walled off as a protoplast at one end of the cell. This core contains the minimum structures and chemicals necessary for guiding life processes. During this time, the sporangium remains active in synthesizing compounds required for spore formation.
3	Sporangium	The protoplast is engulfed by the sporangium to continue the formation of various protective layers around it.
4	Sporangium with prospore	Special peptidoglycan is laid down to form a cortex around the spore protoplast, now called the prospore; calcium and dipicolinic acid are deposited; core becomes dehydrated and metabolically inactive.
5	Sporangium with prospore	Three heavy and impervious protein spore coats are added.
6	Mature endospore	Endospore becomes thicker, and heat resistance is complete; sporangium is no longer functional and begins to deteriorate.
7	Free spore	Complete lysis of sporangium frees spore; it can remain dormant yet viable for thousands of years.
8	Germination	Addition of nutrients and water reverses the dormancy. The spore then swells and liberates a young vegetative cell.
9	Vegetative cell	Restored vegetative cell.

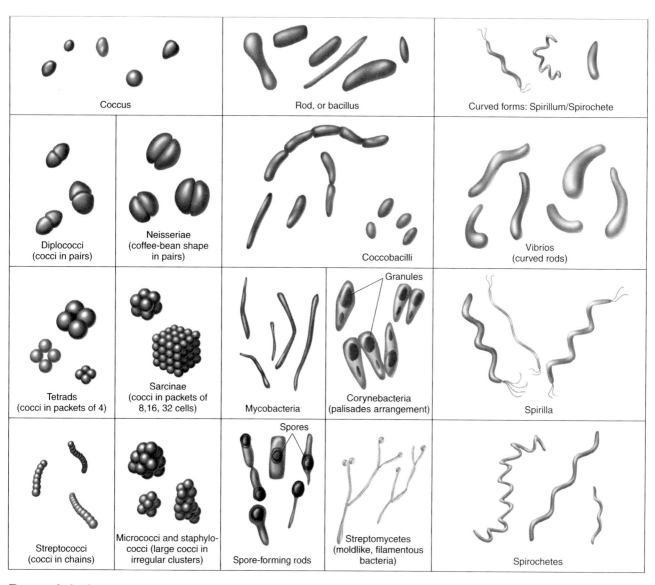

Bacterial shapes and arrangements
Figure 4.22

46

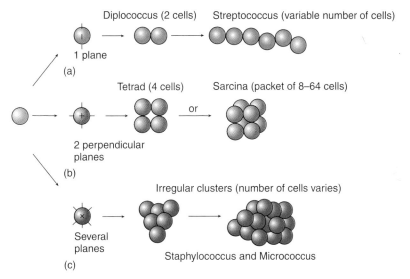

Diplococcus (2 cells)   Streptococcus (variable number of cells)

1 plane

(a)

Tetrad (4 cells)   Sarcina (packet of 8–64 cells)

or

2 perpendicular
planes

(b)

Irregular clusters (number of cells varies)

Several
planes

Staphylococcus and Micrococcus

(c)

Arrangement of cocci resulting from different planes of cell division
Figure 4.25

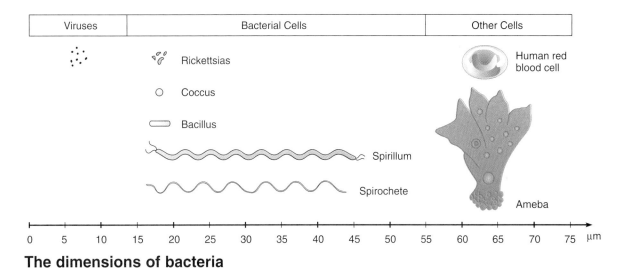

Viruses	Bacterial Cells	Other Cells

Rickettsias

Human red
blood cell

Coccus

Bacillus

Spirillum

Spirochete

Ameba

0   5   10   15   20   25   30   35   40   45   50   55   60   65   70   75   μm

The dimensions of bacteria
Figure 4.26

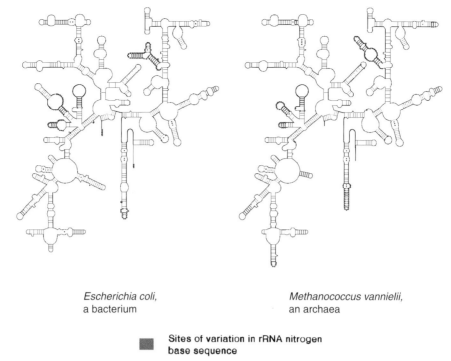

Escherichia coli,
a bacterium

Methanococcus vannielii,
an archaea

■ Sites of variation in rRNA nitrogen base sequence

A modern molecular technique that identifies the cell type or subtype by analyzing the nitrogen base sequence of its ribosomal RNA
Figure 4.28

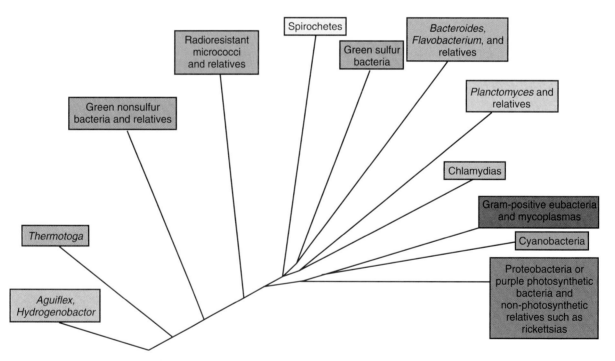

Spirochetes

Radioresistant micrococci and relatives

Green sulfur bacteria

Bacteroides, Flavobacterium, and relatives

Planctomyces and relatives

Green nonsulfur bacteria and relatives

Chlamydias

Gram-positive eubacteria and mycoplasmas

Cyanobacteria

Thermotoga

Aguiflex, Hydrogenobactor

Proteobacteria or purple photosynthetic bacteria and non-photosynthetic relatives such as rickettsias

Separation chart for the eubacteria proposing phylogenetic relationships based on rRNA sequences
Figure 4.30

TABLE 4.6

Medically Important Families and Genera of Bacteria, with Notes on Some Diseases*

I. Bacteria with gram-positive cell wall structure

Cocci in clusters or packets that are aerobic or facultative
    Family Micrococcaceae: *Staphylococcus* (members cause boils, skin infections)

Cocci in pairs and chains that are facultative
    Family Streptococcaceae: *Streptococcus* (species cause strep throat, dental caries)

Anaerobic cocci in pairs, tetrads, irregular clusters
    Family Peptococcaceae: *Peptococcus, Peptostreptococcus* (involved in wound infections)

Spore-forming rods
    Family Bacillaceae: *Bacillus* (anthrax), *Clostridium* (tetanus, gas gangrene, botulism)

Non-spore-forming rods
    Family Lactobacillaceae: *Lactobacillus, Listeria* (milk-borne disease), *Erysipelothrix* (erysipeloid)
    Family Propionibacteriaceae: *Propionibacterium* (involved in acne)

    Family Corynebacteriaceae: *Corynebacterium* (diphtheria)

    Family Mycobacteriaceae: *Mycobacterium* (tuberculosis, leprosy)

    Family Nocardiaceae: *Nocardia* (lung abscesses)

    Family Actinomycetaceae: *Actinomyces* (lumpy jaw), *Bifidobacterium*

    Family Streptomycetaceae: *Streptomyces* (important source of antibiotics)

II. Bacteria with gram-negative cell wall structure

    Family Neisseriaceae
Aerobic cocci
    Neisseria (gonorrhea, meningitis), *Branhamella*
Aerobic coccobacilli
    Moraxella, Acinetobacter
Anaerobic cocci
    Family Veillonellaceae
    Veillonella (dental disease)
Miscellaneous rods
    Brucella (undulant fever), *Bordetella* (whooping cough), *Francisella* (tularemia)
Aerobic rods
    Family Pseudomonadaceae: *Pseudomonas* (pneumonia, burn infections)
    Miscellaneous: *Legionella* (Legionnaires' disease)
Facultative or anaerobic rods and vibrios
    Family Enterobacteriaceae: *Escherichia, Edwardsiella, Citrobacter, Salmonella* (typhoid fever), *Shigella*
    (dysentery), *Klebsiella, Enterobacter, Serratia, Proteus, Yersinia* (one species causes plague)

    Family Vibronaceae: *Vibrio* (cholera, food infection), *Campylobacter, Aeromonas*

    Miscellaneous genera: *Chromobacterium, Flavobacterium, Haemophilus* (meningitis), *Pasteurella,*
    Cardiobacterium, Streptobacillus
Anaerobic rods
    Family Bacteroidaceae: *Bacteroides, Fusobacterium* (anaerobic wound and dental infections)
Helical and curviform bacteria
    Family Spirochaetaceae: *Treponema* (syphilis), *Borrelia* (Lyme disease), *Leptospira* (kidney infection)
Obligate intracellular bacteria
    Family Rickettsiaceae: *Rickettsia* (Rocky Mountain spotted fever), *Coxiella* (Q fever)
    Family Bartonellaceae: *Bartonella* (trench fever, cat scratch disease)
    Family Chlamydiaceae: *Chlamydia* (sexually transmitted infection)

III. Bacteria with no cell walls

    Family Mycoplasmataceae: *Mycoplasma* (pneumonia), *Ureaplasma* (urinary infection)

*Details of pathogens and diseases in chapters 18, 19, 20, and 21.

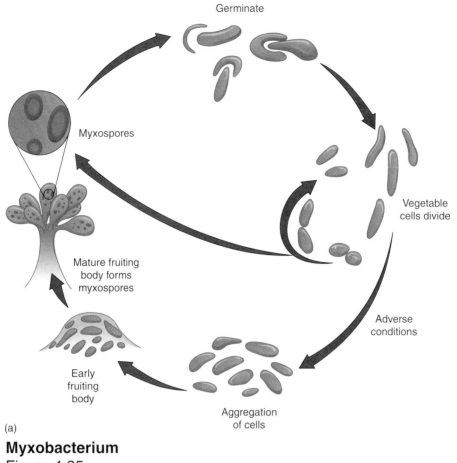

Germinate

Myxospores

Vegetable
cells divide

Mature fruiting
body forms
myxospores

Adverse
conditions

Early
fruiting
body

Aggregation
of cells

(a)

Myxobacterium
Figure 4.35

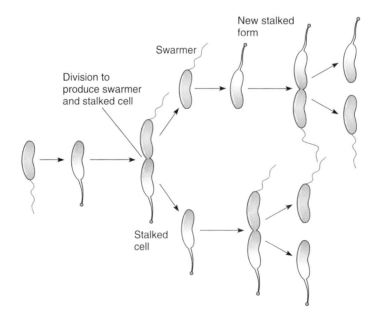

New stalked
form

Swarmer

Division to
produce swarmer
and stalked cell

Stalked
cell

Budding bacteria
Figure 4.36

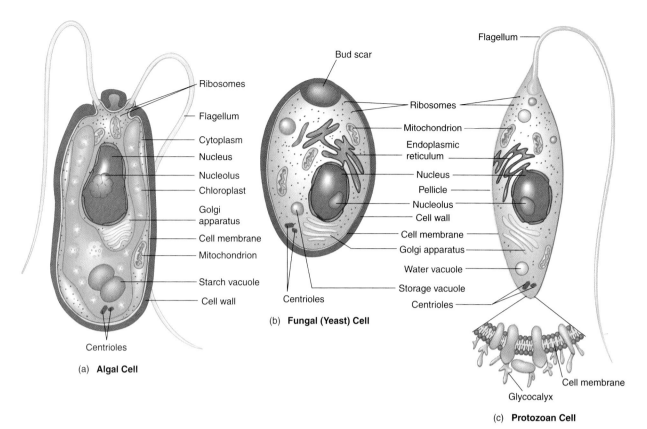

(a) **Algal Cell**

(b) **Fungal (Yeast) Cell**

(c) **Protozoan Cell**

The structure of three representative eucaryotic cells
Figure 5.2

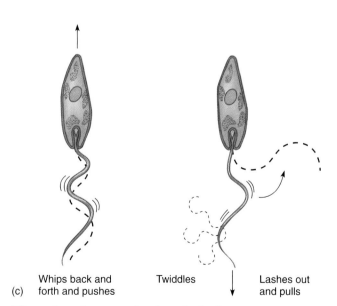

(c) Whips back and forth and pushes

Twiddles

Lashes out and pulls

The structures of microtubules
Figure 5.3

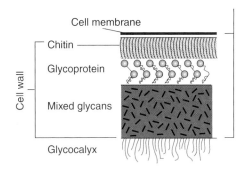

Glycocalyx structure
Figure 5.4

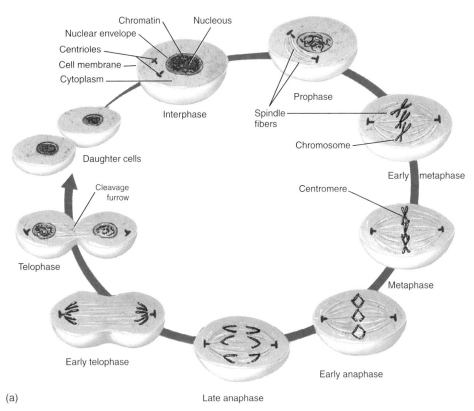

(a)

Changes in the cell and nucleus that accompany mitosis in a eucaryotic cell such as a yeast
Figure 5.6

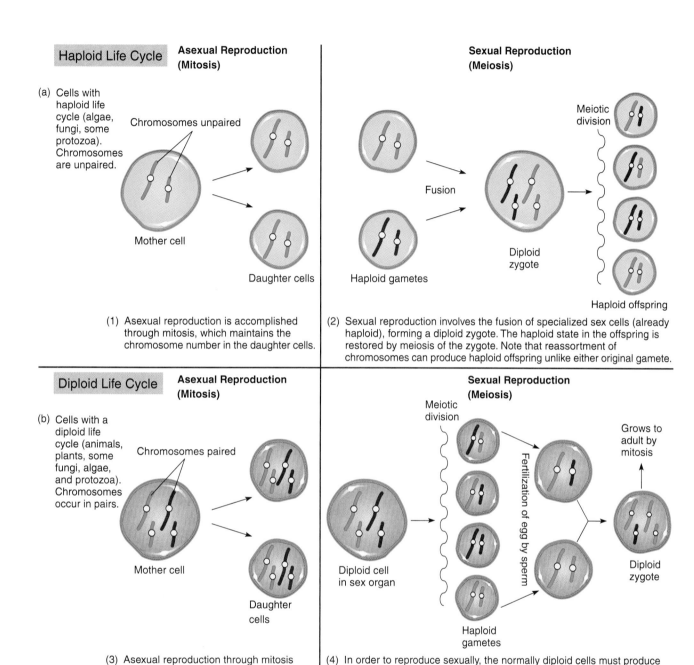

Haploid Life Cycle

Asexual Reproduction (Mitosis)

(a) Cells with haploid life cycle (algae, fungi, some protozoa). Chromosomes are unpaired.

Chromosomes unpaired

Mother cell

Daughter cells

(1) Asexual reproduction is accomplished through mitosis, which maintains the chromosome number in the daughter cells.

Sexual Reproduction (Meiosis)

Fusion

Haploid gametes

Diploid zygote

Meiotic division

Haploid offspring

(2) Sexual reproduction involves the fusion of specialized sex cells (already haploid), forming a diploid zygote. The haploid state in the offspring is restored by meiosis of the zygote. Note that reassortment of chromosomes can produce haploid offspring unlike either original gamete.

Diploid Life Cycle

Asexual Reproduction (Mitosis)

(b) Cells with a diploid life cycle (animals, plants, some fungi, algae, and protozoa). Chromosomes occur in pairs.

Chromosomes paired

Mother cell

Daughter cells

(3) Asexual reproduction through mitosis maintains the chromosome number and content.

Sexual Reproduction (Meiosis)

Meiotic division

Diploid cell in sex organ

Haploid gametes

Fertilization of egg by sperm

Grows to adult by mitosis

Diploid zygote

(4) In order to reproduce sexually, the normally diploid cells must produce haploid gametes through meiotic division. During fertilization, these haploid gametes (egg, sperm) form a zygote that restores the diploid number. Depending upon which gametes fuse, combinations of chromosomes unlike the original parents are possible in offspring.

Schematic of haploid versus diploid life cycles
Figure 5.7

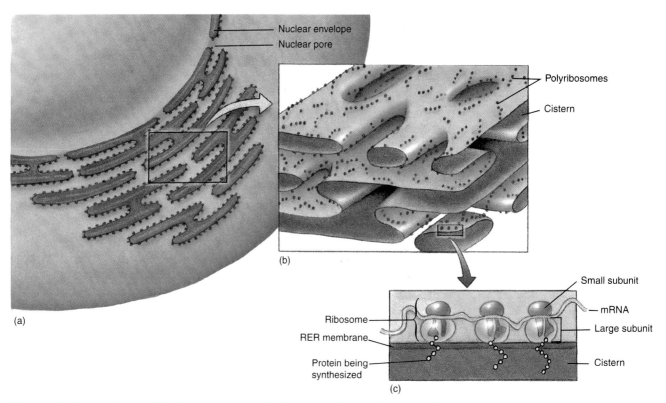

Nuclear envelope
Nuclear pore
Polyribosomes
Cistern
(a)
(b)

Small subunit
mRNA
Ribosome
Large subunit
RER membrane
Cistern
Protein being synthesized
(c)

The origin and detailed structure of the rough endoplasmic reticulum (RER)
Figure 5.8

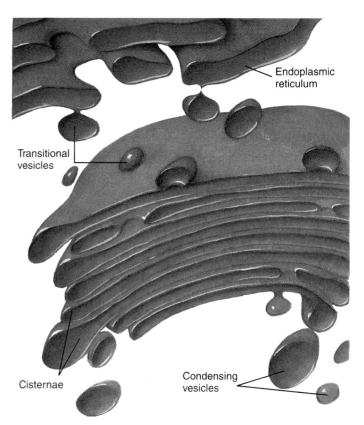

Endoplasmic reticulum
Transitional vesicles
Cisternae
Condensing vesicles

Detail of the Golgi apparatus
Figure 5.9

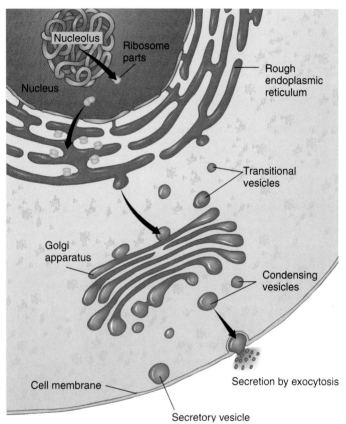

The transport process
Figure 5.10

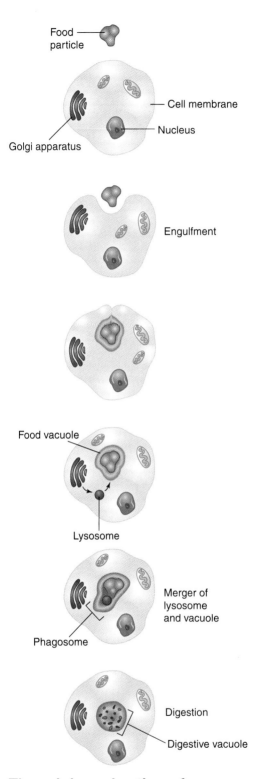

Food particle

Cell membrane

Nucleus

Golgi apparatus

Engulfment

Food vacuole

Lysosome

Merger of lysosome and vacuole

Phagosome

Digestion

Digestive vacuole

The origin and action of lysosomes in phagocytosis
Figure 5.11

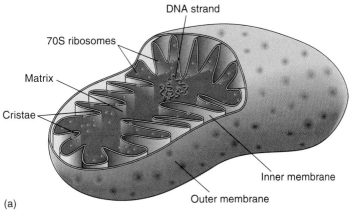

(a)

General structure of a mitochondrion
Figure 5.12

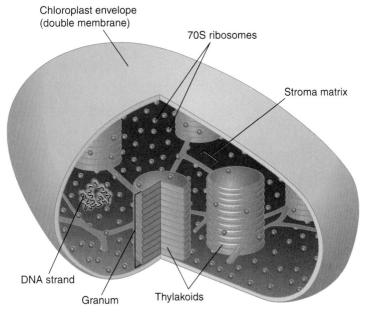

Detail of an algal chloroplast
Figure 5.13

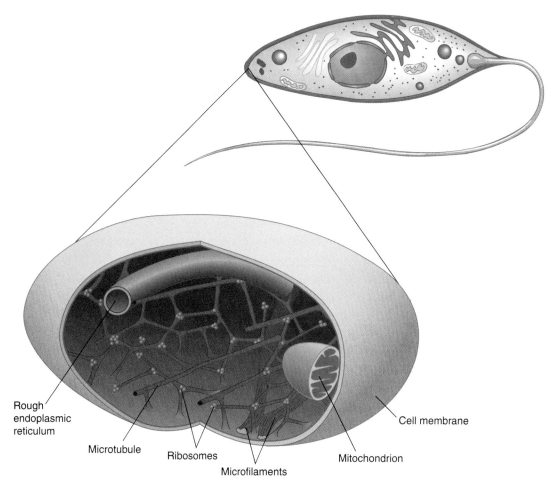

Rough
endoplasmic
reticulum

Microtubule

Ribosomes

Microfilaments

Mitochondrion

Cell membrane

A model of the cytoskeleton
Figure 5.14

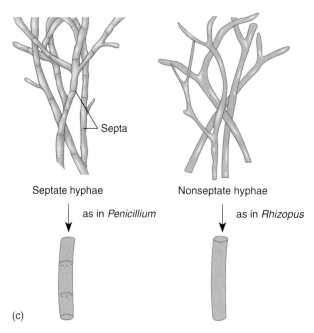

Septa

Septate hyphae

as in *Penicillium*

Nonseptate hyphae

as in *Rhizopus*

(c)

***Diplodia maydis,* a pathogenic fungus
of corn plants**
Figure 5.15

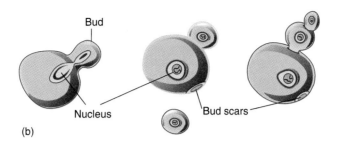

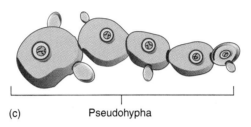

Microscopic morphology of yeasts
Figure 5.16

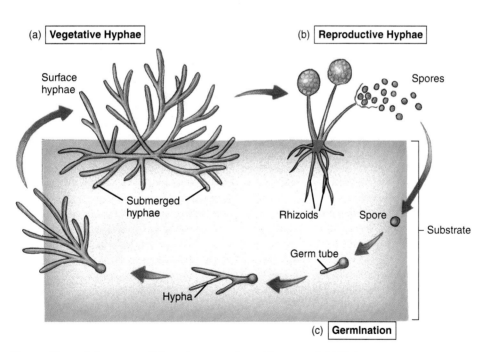

Functional types of hyphae using the mold Rhizopus as an example
Figure 5.18

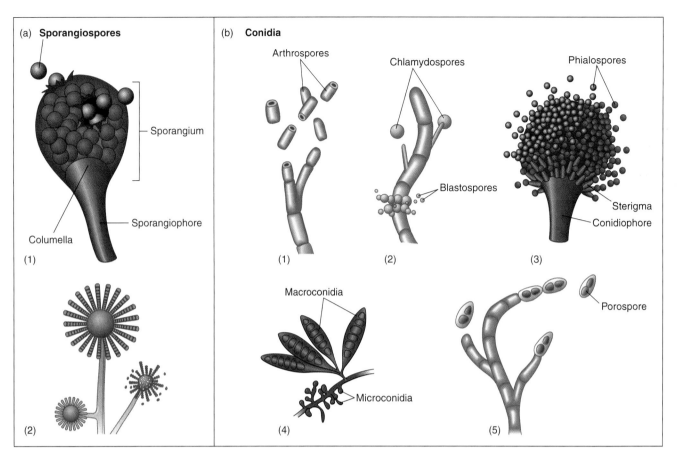

(a) **Sporangiospores**

Sporangium

Sporangiophore

Columella

Sporangiophore

(1)

(2)

(b) **Conidia**

Arthrospores

Chlamydospores

Phialospores

Blastospores

Sterigma

Conidiophore

(1)

(2)

(3)

Macroconidia

Microconidia

Porospore

(4)

(5)

Types of asexual mold spores
Figure 5.19

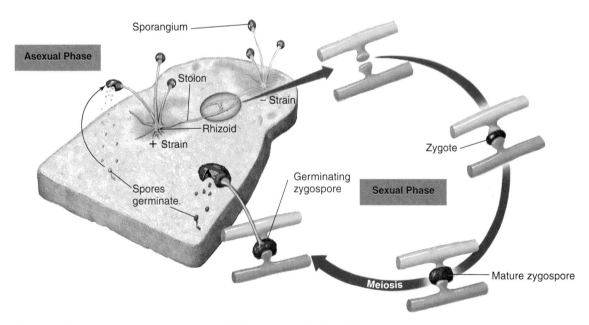

Asexual Phase

Sporangium

Stolon

− Strain

Rhizoid

+ Strain

Spores germinate.

Zygote

Germinating zygospore

Sexual Phase

Meiosis

Mature zygospore

Formation of zygospores in *Rhizopus stolonifer*
Figure 5.20

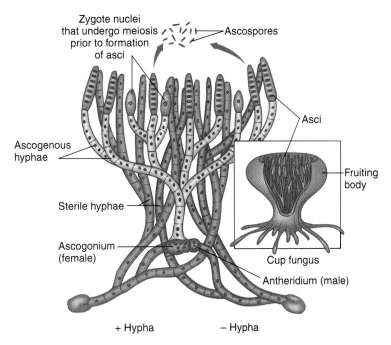

Production of ascospores in a cup fungus
Figure 5.21

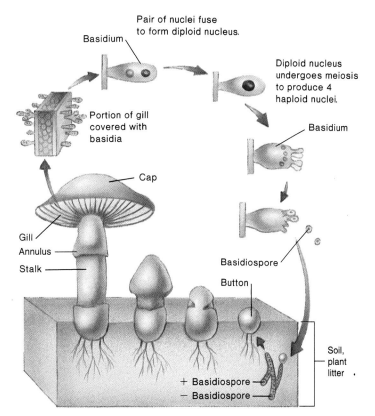

Formation of basidiospores in a mushroom
Figure 5.22

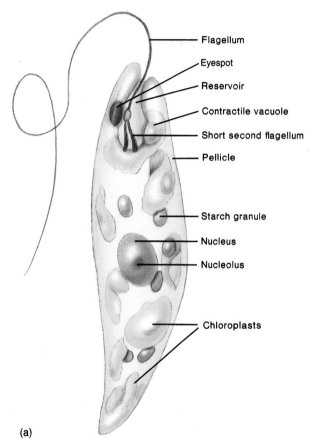

Flagellum

Eyespot

Reservoir

Contractile vacuole

Short second flagellum

Pellicle

Starch granule

Nucleus

Nucleolus

Chloroplasts

(a)

Representative microscopic algae
Figure 5.26

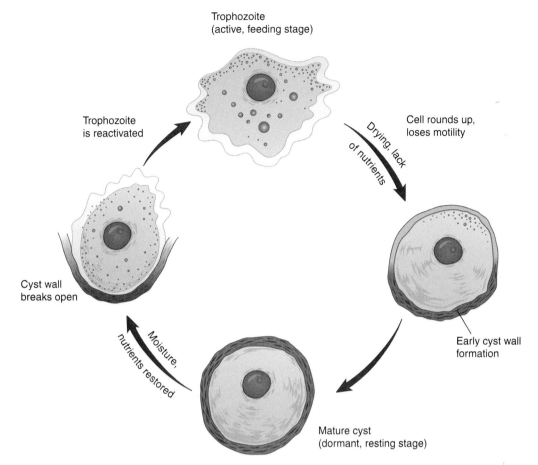

Trophozoite
(active, feeding stage)

Trophozoite
is reactivated

Cell rounds up,
loses motility

Drying. lack
of nutrients

Cyst wall
breaks open

Early cyst wall
formation

Moisture,
nutrients restored

Mature cyst
(dormant, resting stage)

The general life cycle exhibited by many protozoa
Figure 5.27

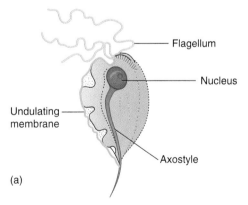

Flagellum

Nucleus

Undulating
membrane

Axostyle

(a)

**The structure of a typical
mastigophoran, *Trichomonas
vaginalis***
Figure 5.28

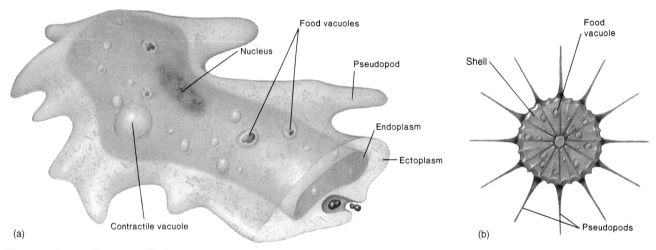

Examples of sarcodinians
Figure 5.29

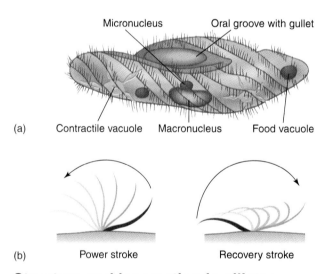

Structure and locomotion in ciliates
Figure 5.30

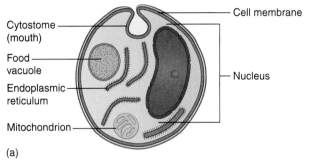

Cytostome (mouth)
Food vacuole
Endoplasmic reticulum
Mitochondrion
Cell membrane
Nucleus

(a)

Sporozoan parasites
Figure 5.32

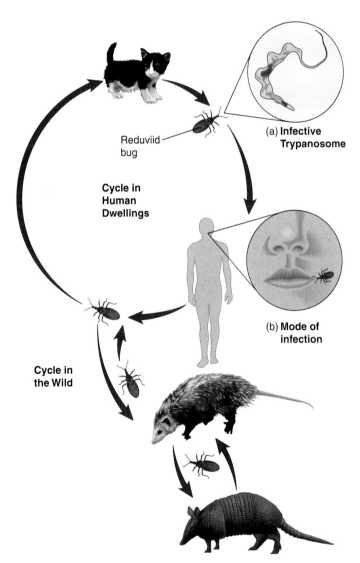

Reduviid bug
(a) **Infective Trypanosome**

Cycle in Human Dwellings

(b) **Mode of infection**

Cycle in the Wild

Cycle of transmission in Chagas disease
Figure 5.33

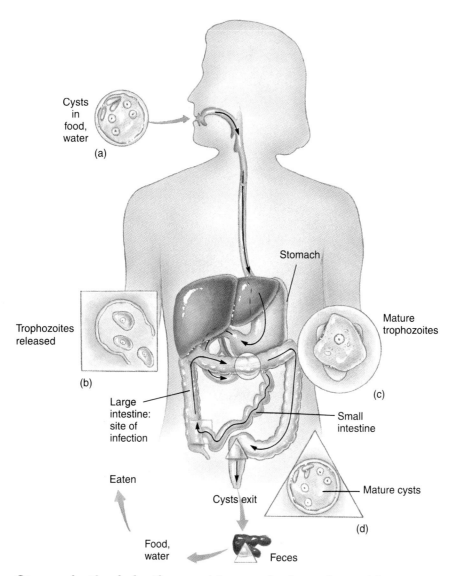

Cysts
in
food,
water
(a)

Trophozoites
released
(b)

Stomach

Mature
trophozoites
(c)

Large
intestine:
site of
infection

Small
intestine

Mature cysts
(d)

Eaten

Cysts exit

Food,
water

Feces

Stages in the infection and transmission of amebic dysentery
Figure 5.34

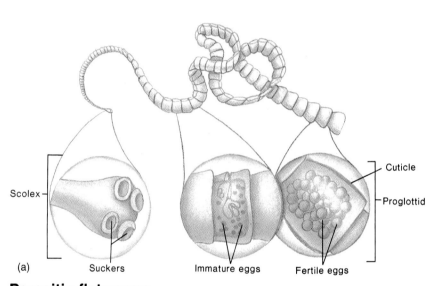

(a)

Scolex

Suckers

Immature eggs

Fertile eggs

Cuticle

Proglottid

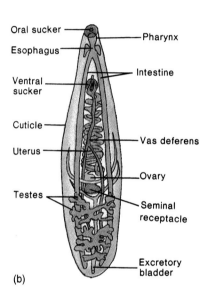

Oral sucker

Esophagus

Ventral sucker

Cuticle

Uterus

Testes

Pharynx

Intestine

Vas deferens

Ovary

Seminal receptacle

Excretory bladder

(b)

Parasitic flatworms
Figure 5.35

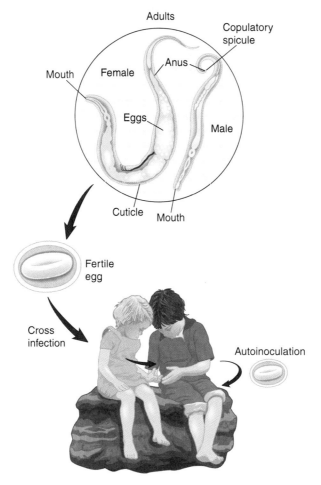

The life cycle of the pinworm, a roundworm
Figure 5.36

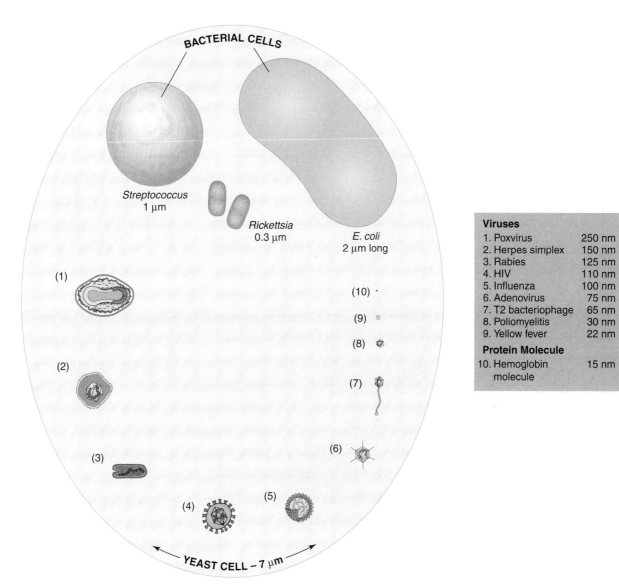

Size comparison of viruses with a eucaryotic cell (yeast) and bacteria
Figure 6.1

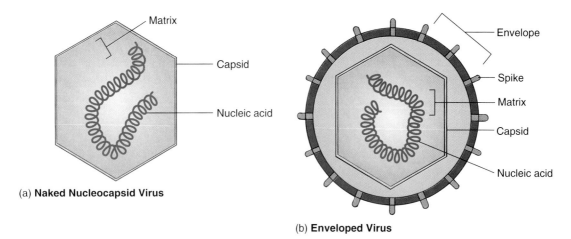

(a) **Naked Nucleocapsid Virus**

(b) **Enveloped Virus**

Generalized structure of viruses
Figure 6.4

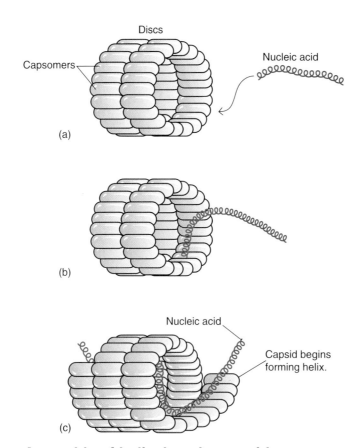

Assembly of helical nucleocapsids
Figure 6.5

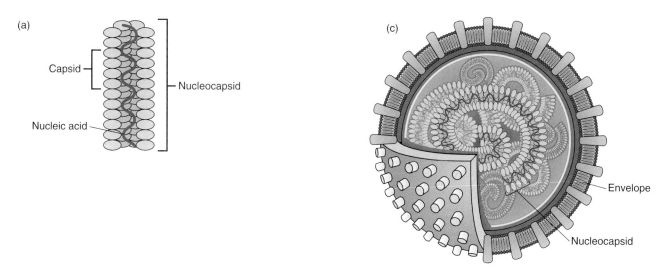

Typical variations of viruses with helical nucleocapsids
Figure 6.6

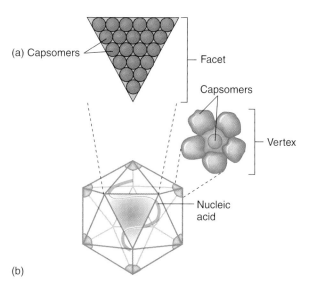

The structure and formation of an icosahedral virus (adenovirus is the model)
Figure 6.7

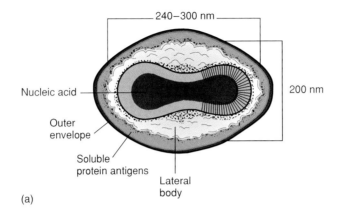

240–300 nm

200 nm

Nucleic acid

Outer
envelope

Soluble
protein antigens

Lateral
body

(a)

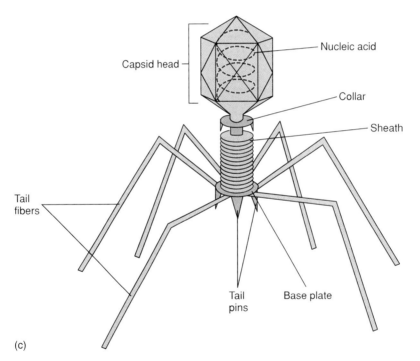

Capsid head

Nucleic acid

Collar

Sheath

Tail
fibers

Tail
pins

Base plate

(c)

Detailed structure of complex viruses
Figure 6.9

a: From Westwood, et al., *Journal of Microbiology,* 34:67, 1964. Reprinted by permission of The Society for General Microbiology, United Kingdom.

Complex Viruses | Enveloped Viruses | Naked Viruses

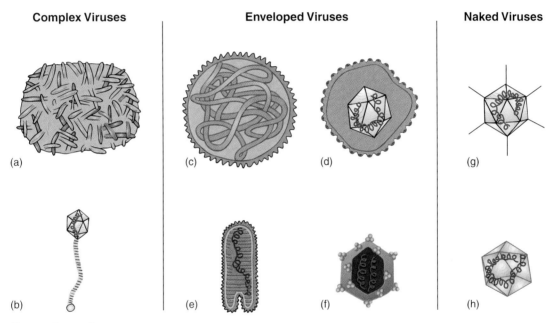

(a)

(b)

(c)

(d)

(e)

(f)

(g)

(h)

Complex viruses
Figure 6.10

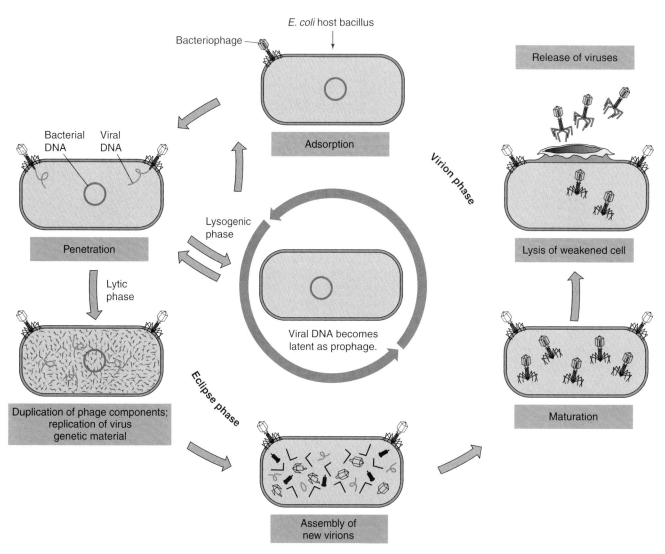

E. coli host bacillus

Bacteriophage

Adsorption

Bacterial DNA   Viral DNA

Penetration

Lytic phase

Lysogenic phase

Viral DNA becomes latent as prophage.

Virion phase

Release of viruses

Lysis of weakened cell

Duplication of phage components; replication of virus genetic material

Eclipse phase

Assembly of new virions

Maturation

Events in the multiplication cycle of T-even bacteriophages
Figure 6.11

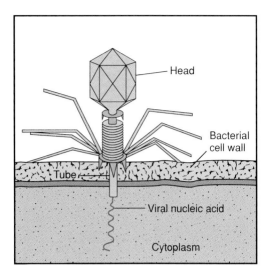

**Penetration of a bacterial cell by a
T-even bacteriophage**
Figure 6.12

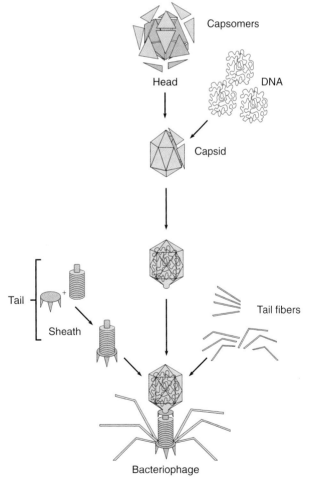

Bacteriophage assembly line
Figure 6.13

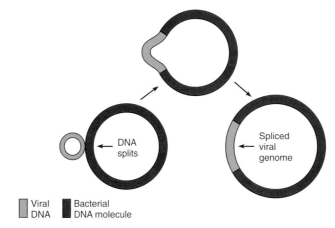

Viral
DNA

Bacterial
DNA molecule

DNA
splits

Spliced
viral
genome

The lysogenic state in bacteria
Figure 6.15

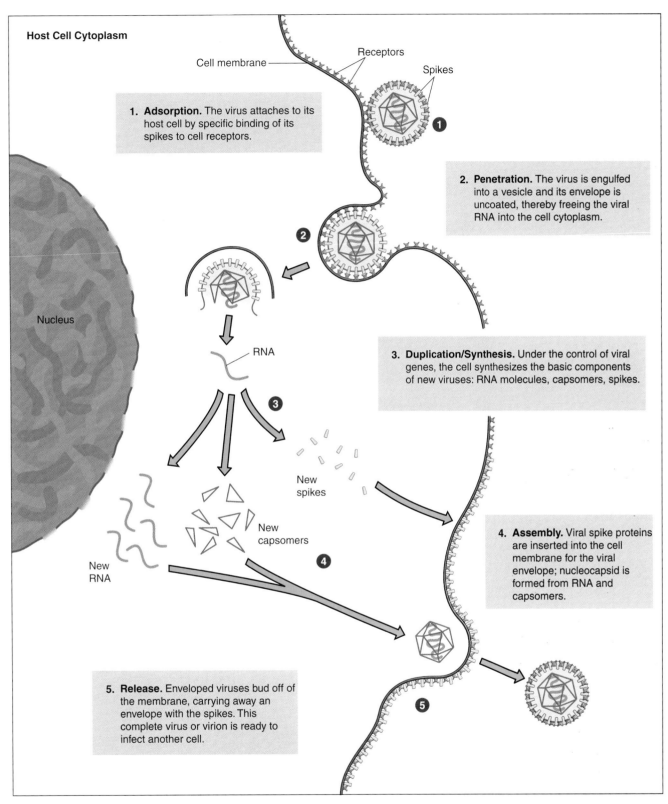

Host Cell Cytoplasm

Cell membrane

Receptors

Spikes

1. Adsorption. The virus attaches to its host cell by specific binding of its spikes to cell receptors.

❶

2. Penetration. The virus is engulfed into a vesicle and its envelope is uncoated, thereby freeing the viral RNA into the cell cytoplasm.

❷

Nucleus

RNA

❸

3. Duplication/Synthesis. Under the control of viral genes, the cell synthesizes the basic components of new viruses: RNA molecules, capsomers, spikes.

New spikes

New capsomers

❹

New RNA

4. Assembly. Viral spike proteins are inserted into the cell membrane for the viral envelope; nucleocapsid is formed from RNA and capsomers.

5. Release. Enveloped viruses bud off of the membrane, carrying away an envelope with the spikes. This complete virus or virion is ready to infect another cell.

❺

General features in the multiplication cycle of an enveloped animal virus
Figure 6.16

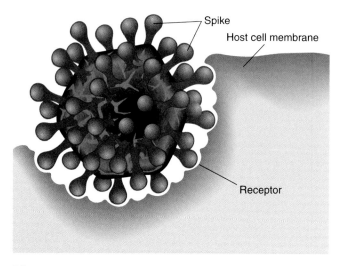

Spike

Host cell membrane

Receptor

(a)

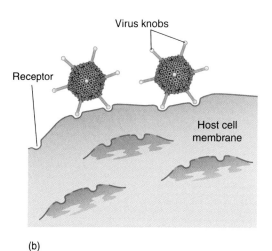

Virus knobs

Receptor

Host cell membrane

(b)

The mode by which animal viruses adsorb to the host cell membrane
Figure 6.17

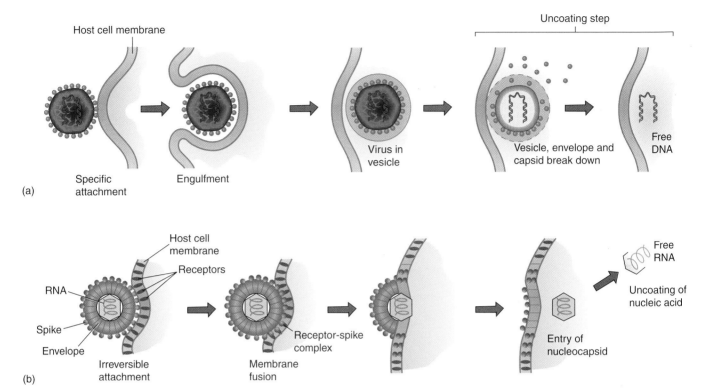

(a) Specific attachment → Engulfment → Virus in vesicle → Vesicle, envelope and capsid break down → Free DNA

Host cell membrane

Uncoating step

(b) Host cell membrane — Receptors — RNA — Spike — Envelope — Irreversible attachment → Receptor-spike complex / Membrane fusion → Entry of nucleocapsid → Free RNA / Uncoating of nucleic acid

Two principal means by which animal viruses penetrate
Figure 6.18

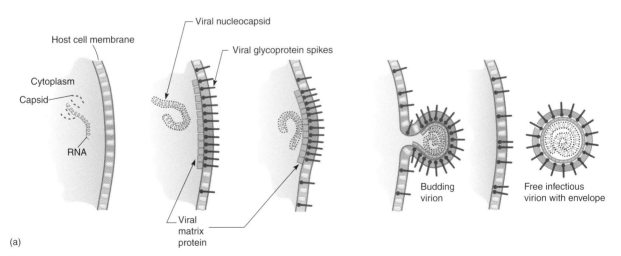

Host cell membrane — Viral nucleocapsid — Viral glycoprotein spikes — Cytoplasm — Capsid — RNA — Viral matrix protein — Budding virion — Free infectious virion with envelope

(a)

Process of maturation of an enveloped virus (parainfluenza virus)
Figure 6.20

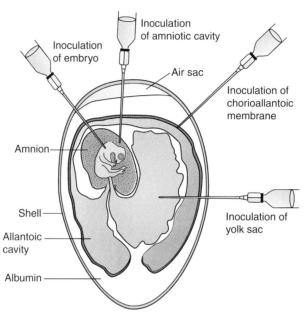

(b) The shell is perforated using sterile techniques, and a virus preparation is injected into a site selected to grow the viruses. Targets include the allantoic cavity, a fluid-filled sac that functions in embryonic waste removal; the amniotic cavity, a sac that cushions and protects the embryo itself; the chorioallantoic membrane, which functions in embryonic gas exchange; the yolk sac, a membrane that mobilizes yolk for the nourishment of the embryo; and the embryo itself.

Cultivating animal viruses in a developing bird embryo
Figure 6.22

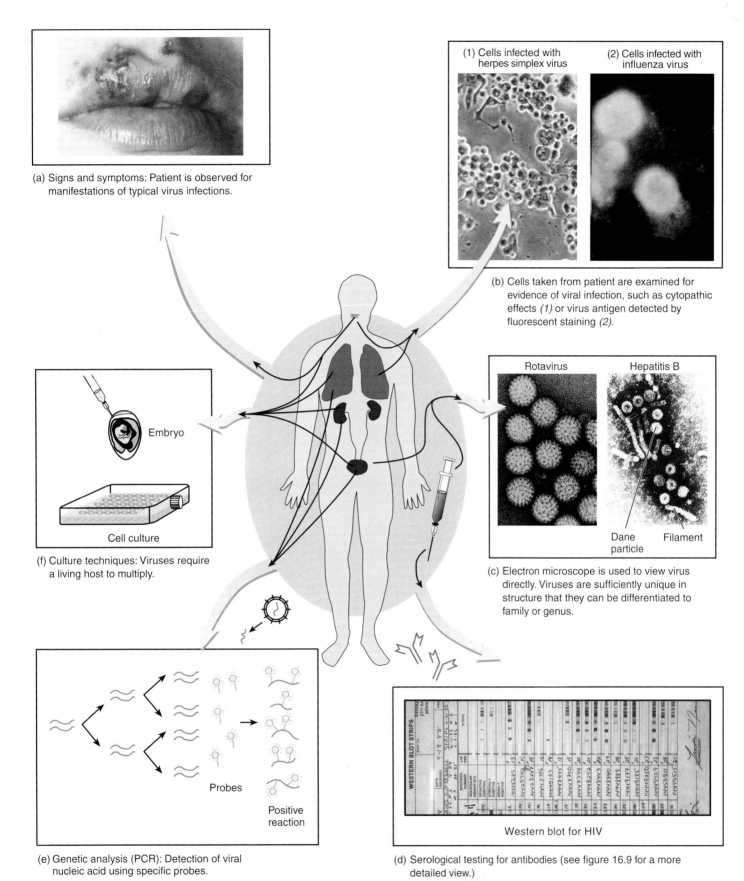

(a) Signs and symptoms: Patient is observed for manifestations of typical virus infections.

(1) Cells infected with herpes simplex virus

(2) Cells infected with influenza virus

(b) Cells taken from patient are examined for evidence of viral infection, such as cytopathic effects *(1)* or virus antigen detected by fluorescent staining *(2)*.

Rotavirus

Hepatitis B

Dane particle

Filament

(c) Electron microscope is used to view virus directly. Viruses are sufficiently unique in structure that they can be differentiated to family or genus.

Embryo

Cell culture

(f) Culture techniques: Viruses require a living host to multiply.

Probes

Positive reaction

(e) Genetic analysis (PCR): Detection of viral nucleic acid using specific probes.

Western blot for HIV

(d) Serological testing for antibodies (see figure 16.9 for a more detailed view.)

Summary of methods used to diagnose viral infections
Figure 6.24

a: Carroll M. Weiss/Camera M.D. Studios; **b1:** © A.M. Siegelman/Visuals Unlimited; **b2:** © Science VU/CDC/Visuals Unlimited; **c1:** © K.G. Murti/Visuals Unlimited; **c2:** Courtesy Fred Williams, U.S. Environmental Protection Agency; **d:** Courtesy of Reference Laboratory, A PCC Laboratory.

Digestion in Bacteria and Fungi

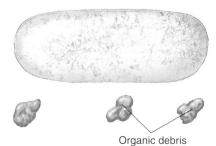

Organic debris

(a)   Walled cell is inflexible.

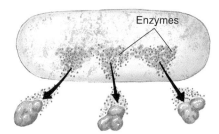

Enzymes

(b)   Enzymes are transported across the wall.

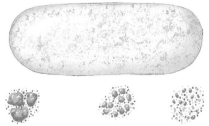

(c)   Enzymes hydrolyze the bonds on nutrients.

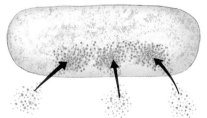

(d)   Smaller molecules are transported into the cytoplasm.

Extracellular digestion in a saprobe with a cell wall (bacterium or fungus)

Figure 7.2

How Molecules Diffuse In Aqueous Solutions

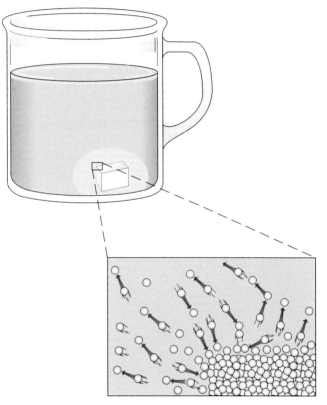

Diffusion of molecules in aqueous solutions
Figure 7.3

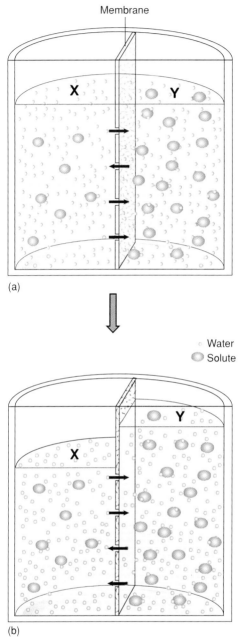

Membrane

X   Y

(a)

○ Water
◎ Solute

X   Y

(b)

Osmosis, the diffusion of water through a selectively permeable membrane
Figure 7.4

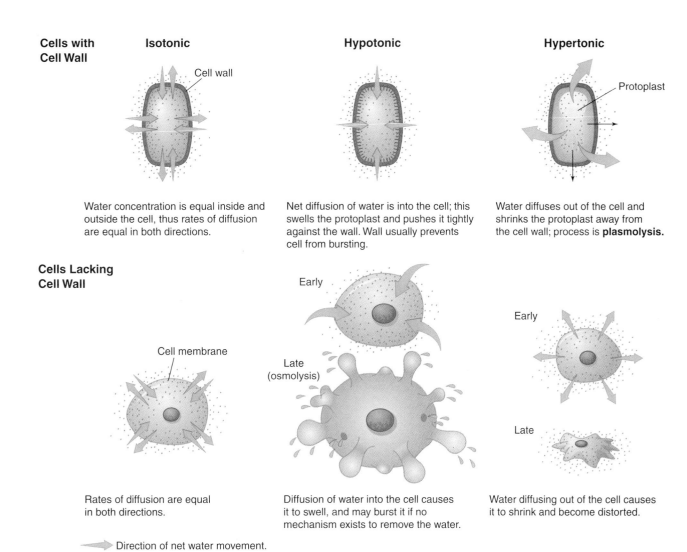

Cells with Cell Wall

Isotonic

Cell wall

Water concentration is equal inside and outside the cell, thus rates of diffusion are equal in both directions.

Hypotonic

Net diffusion of water is into the cell; this swells the protoplast and pushes it tightly against the wall. Wall usually prevents cell from bursting.

Hypertonic

Protoplast

Water diffuses out of the cell and shrinks the protoplast away from the cell wall; process is **plasmolysis.**

Cells Lacking Cell Wall

Cell membrane

Rates of diffusion are equal in both directions.

Early

Late (osmolysis)

Diffusion of water into the cell causes it to swell, and may burst it if no mechanism exists to remove the water.

Early

Late

Water diffusing out of the cell causes it to shrink and become distorted.

Direction of net water movement.

Cell responses to solutions of differing osmotic content
Figure 7.5

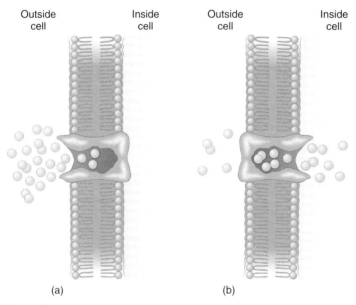

Outside   Inside
cell   cell

Outside   Inside
cell   cell

(a)   (b)

Passive transport
Figure 7.6

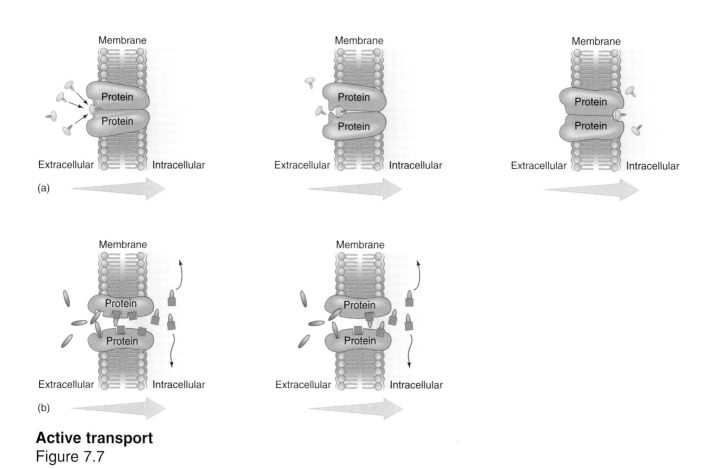

Active transport
Figure 7.7

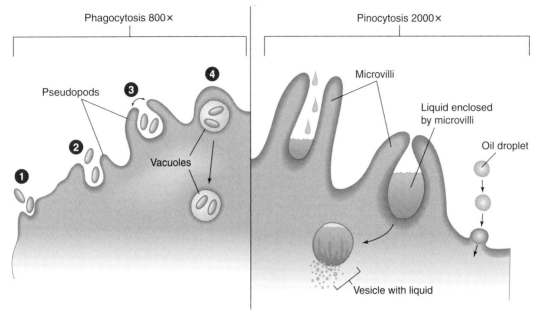

Phagocytosis 800×

Pinocytosis 2000×

Pseudopods

Vacuoles

Microvilli

Liquid enclosed
by microvilli

Oil droplet

Vesicle with liquid

Endocytosis (phagocytosis and pinocytosis)
Figure 7.8

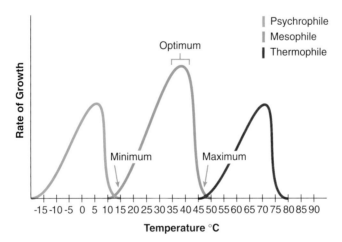

Psychrophile
Mesophile
Thermophile

Optimum

Rate of Growth

Minimum

Maximum

-15 -10 -5 0 5 10 15 20 25 30 35 40 45 50 55 60 65 70 75 80 85 90
Temperature °C

**Ecological groups by temperature of
adaptation**
Figure 7.9

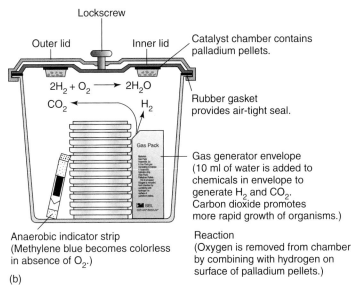

Lockscrew

Outer lid

Inner lid

Catalyst chamber contains palladium pellets.

$$2H_2 + O_2 \longrightarrow 2H_2O$$

CO_2 H_2

Rubber gasket provides air-tight seal.

Gas Pack

Gas generator envelope (10 ml of water is added to chemicals in envelope to generate H_2 and CO_2. Carbon dioxide promotes more rapid growth of organisms.)

Anaerobic indicator strip (Methylene blue becomes colorless in absence of O_2.)

Reaction (Oxygen is removed from chamber by combining with hydrogen on surface of palladium pellets.)

(b)

Culturing techniques for anaerobes
Figure 7.11

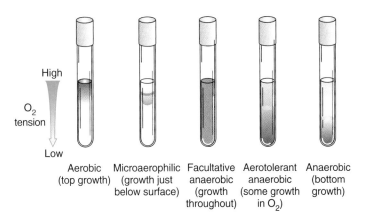

Demonstration of Oxygen Requirements

High

O_2 tension

Low

Aerobic (top growth)

Microaerophilic (growth just below surface)

Facultative anaerobic (growth throughout)

Aerotolerant anaerobic (some growth in O_2)

Anaerobic (bottom growth)

Use of thioglycollate broth to demonstrate oxygen requirements
Figure 7.12

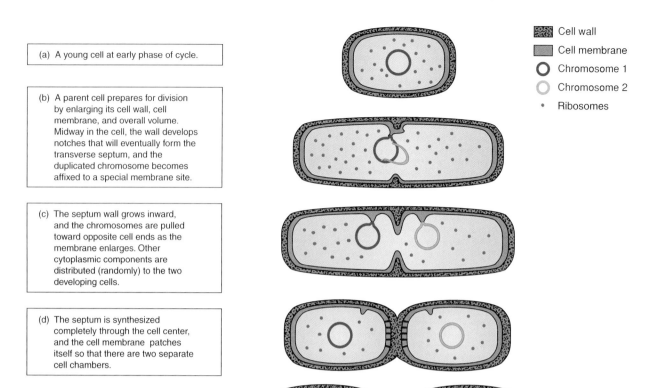

(a) A young cell at early phase of cycle.

(b) A parent cell prepares for division by enlarging its cell wall, cell membrane, and overall volume. Midway in the cell, the wall develops notches that will eventually form the transverse septum, and the duplicated chromosome becomes affixed to a special membrane site.

(c) The septum wall grows inward, and the chromosomes are pulled toward opposite cell ends as the membrane enlarges. Other cytoplasmic components are distributed (randomly) to the two developing cells.

(d) The septum is synthesized completely through the cell center, and the cell membrane patches itself so that there are two separate cell chambers.

(e) At this point, the daughter cells are divided. Some species will separate completely as shown here, while others will remain attached.

	Cell wall
	Cell membrane
O	Chromosome 1
O	Chromosome 2
•	Ribosomes

Steps in binary fission of a rod-shaped bacterium
Figure 7.14

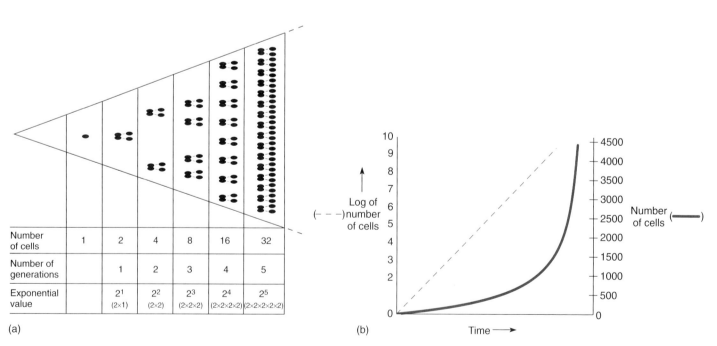

Number of cells	1	2	4	8	16	32
Number of generations		1	2	3	4	5
Exponential value		2^1 (2×1)	2^2 (2×2)	2^3 (2×2×2)	2^4 (2×2×2×2)	2^5 (2×2×2×2×2)

(a)

(b)

The mathematics of population growth
Figure 7.15

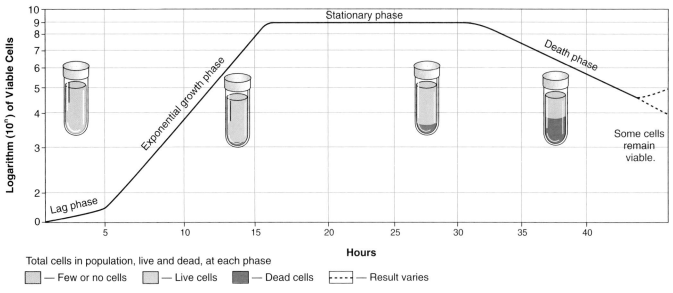

The growth curve in a bacterial culture
Figure 7.16

Total cells in population, live and dead, at each phase

☐ — Few or no cells   ☐ — Live cells   ■ — Dead cells   ┆- - -┆ — Result varies

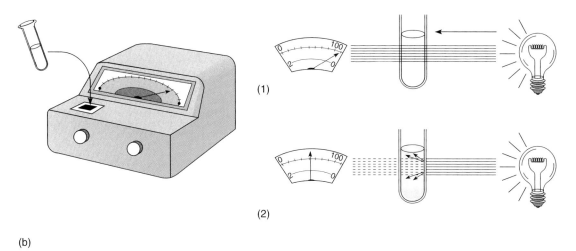

(b)

Turbidity measurements as indicators of growth
Figure 7.17

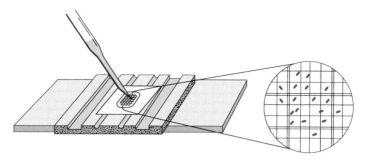

Direct microscopic count of bacteria
Figure 7.18

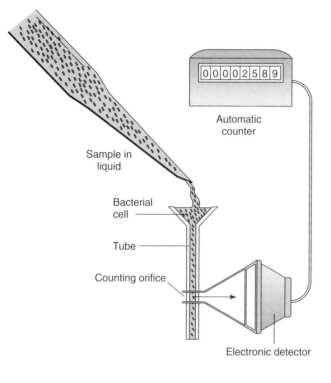

(a) **Coulter counter.** As cells pass through this device, they trigger an electronic sensor that tallies their numbers.

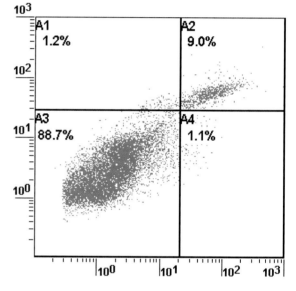

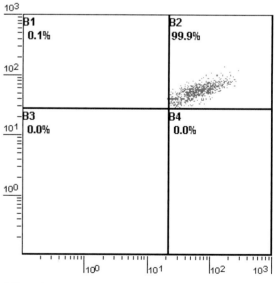

(b) **Flow cytometry.** A variation on the Coulter counter is used to detect, count, and sort the cells in a sample and display a plot of the results. In one version of this procedure, zoospores of the toxic algae *Pfiesteria* (B$_2$) have been separated from a mixed sample (A$_3$). The method is so accurate and precise that the sorted sample is a pure culture of the zoospores.

Methods of electronic counting, sorting, and identification
Figure 7.19

b: Image courtesy Center for Applied Aquatic Ecology

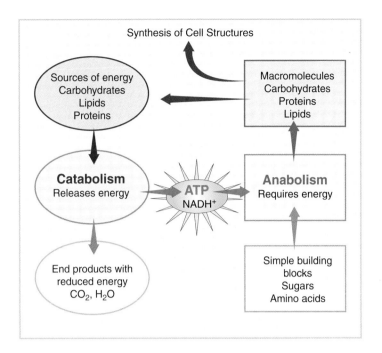

Simplified Model of Metabolism
Figure 8.1

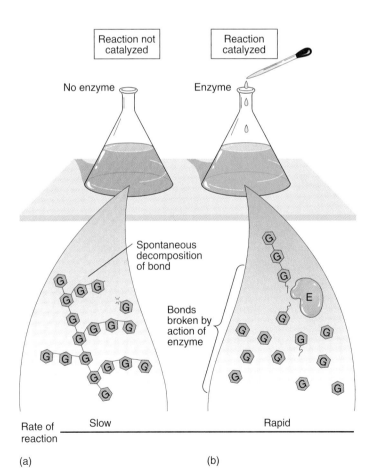

The effects of a catalyst
Figure 8.2

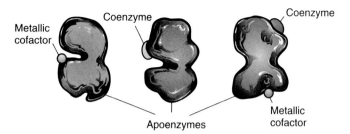

Conjugated enzyme structure
Figure 8.3

Levels of Structure

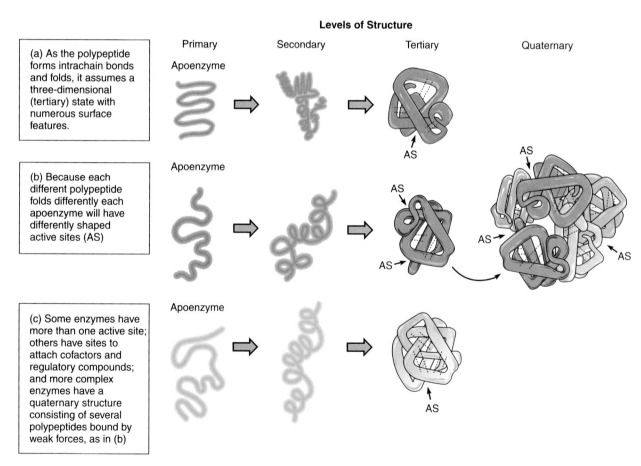

Primary   Secondary   Tertiary   Quaternary

(a) As the polypeptide forms intrachain bonds and folds, it assumes a three-dimensional (tertiary) state with numerous surface features.

Apoenzyme

AS

(b) Because each different polypeptide folds differently each apoenzyme will have differently shaped active sites (AS)

Apoenzyme

AS

AS

AS

AS

AS

AS

(c) Some enzymes have more than one active site; others have sites to attach cofactors and regulatory compounds; and more complex enzymes have a quaternary structure consisting of several polypeptides bound by weak forces, as in (b)

Apoenzyme

AS

How the active site and specificity of the apoenzyme arise
Figure 8.4

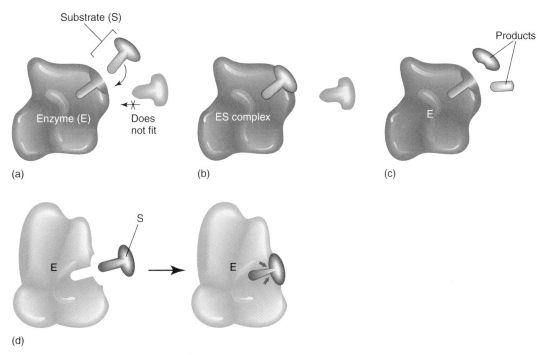

Enzyme-substrate reactions
Figure 8.5

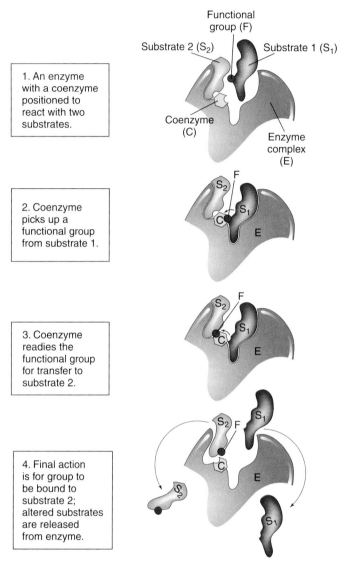

1. An enzyme with a coenzyme positioned to react with two substrates.

2. Coenzyme picks up a functional group from substrate 1.

3. Coenzyme readies the functional group for transfer to substrate 2.

4. Final action is for group to be bound to substrate 2; altered substrates are released from enzyme.

Functional group (F)

Substrate 2 (S_2)

Substrate 1 (S_1)

Coenzyme (C)

Enzyme complex (E)

The carrier functions of coenzymes
Figure 8.6

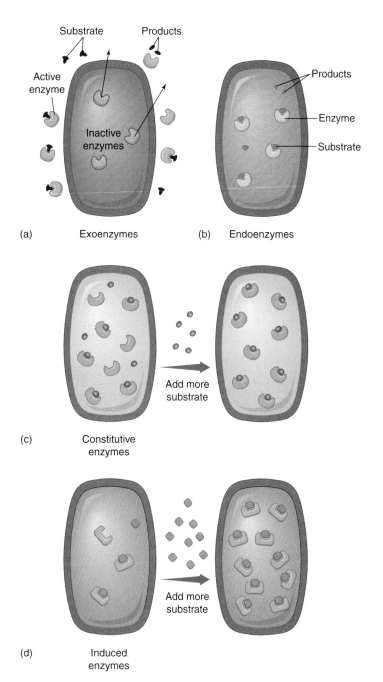

(a)   Exoenzymes   (b)   Endoenzymes

(c)   Constitutive enzymes

(d)   Induced enzymes

Types of enzymes, as described by their location of action and quantity

Figure 8.7

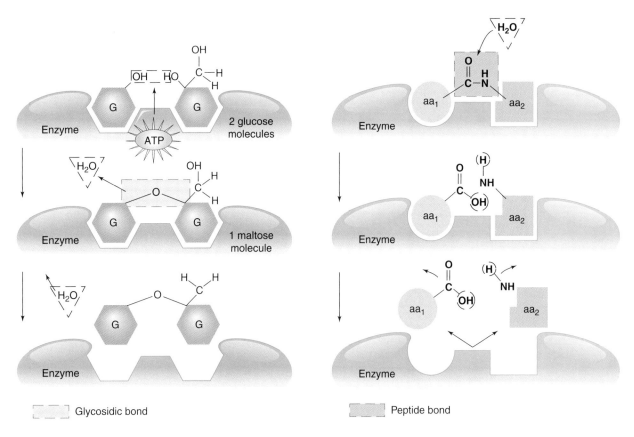

Glycosidic bond

Peptide bond

(a) **Condensation Reaction.** Forming a glycosidic bond between two glucose molecules to generate maltose requires the removal of a water molecule and energy from ATP.

(b) **Hydrolysis Reaction.** Breaking a peptide bond between two amino acids requires a water molecule that adds an H and OH to the amino acids.

Examples of enzyme-catalyzed synthesis and hydrolysis reactions
Figure 8.8

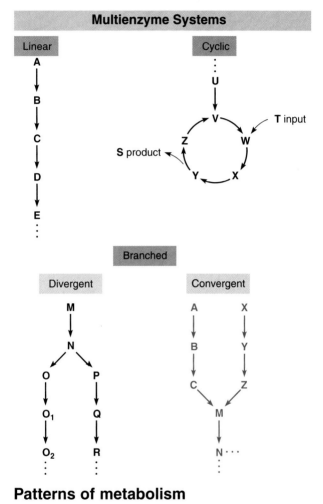

Patterns of metabolism
Figure 8.9

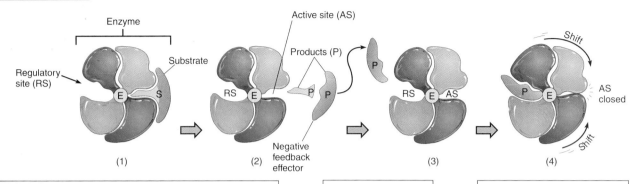

(1), (2) Allosteric enzymes have a quaternary structure with two different sites of attachment—the active site (AS) and the regulatory site (RS). The enzyme complex normally attaches to the substrate at the AS and releases products (P).

(3) One product can function as a negative-feedback effector by fitting into a regulatory site.

(4) The entrance of P into RS causes a confirmational shift of the enzyme that closes the active site. The enzyme cannot catalyze further reactions with substrate as long as the product is inserted in the regulatory site.

Competitive Inhibition

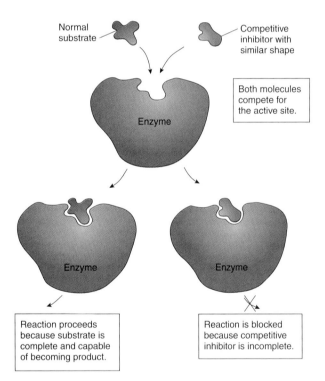

Both molecules compete for the active site.

Reaction proceeds because substrate is complete and capable of becoming product.

Reaction is blocked because competitive inhibitor is incomplete.

Examples of two common control mechanisms for enzymes
Figure 8.10

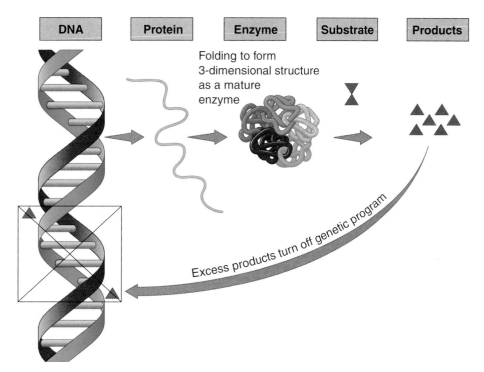

| DNA | Protein | Enzyme | Substrate | Products |

Folding to form
3-dimensional structure
as a mature
enzyme

Excess products turn off genetic program

One type of genetic control of enzyme synthesis: feedback/ enzyme repression
Figure 8.11

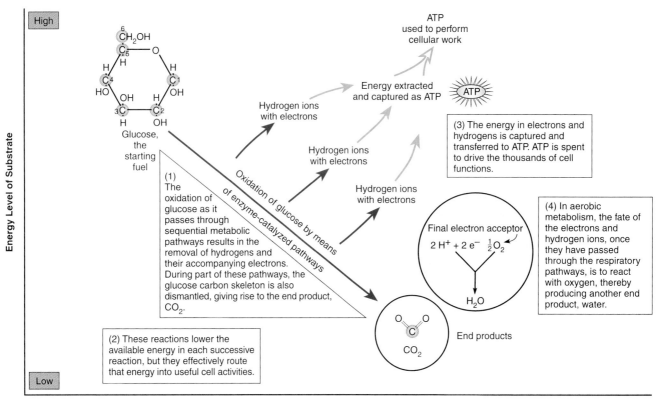

A simplified model that summarizes the cell's energy machine
Figure 8.12

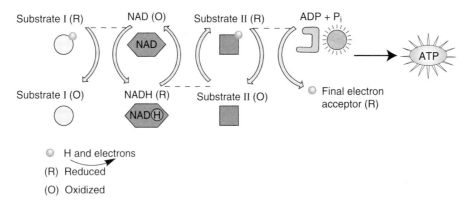

The role of electron carriers in redox reactions
Figure 8.13

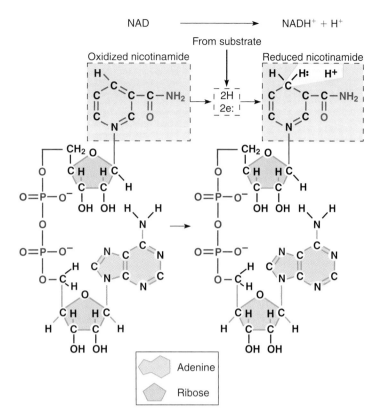

NAD ⟶ NADH⁺ + H⁺

Details of NAD reduction
Figure 8.14

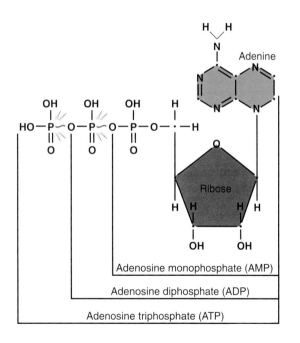

⁀ Bond that releases
energy when broken

The structure of adenosine triphosphate (ATP) and its partner compounds, ADP and AMP
Figure 8.15

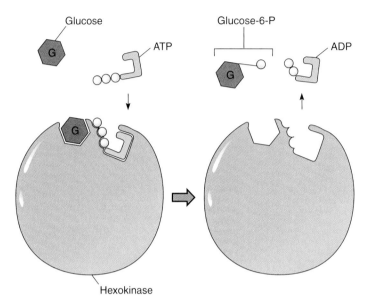

An example of phosphorylation of glucose by ATP
Figure 8.16

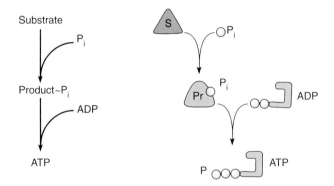

ATP formation at the substrate level shown in outline and illustrated form
Figure 8.17

Location in cell	Pathway involved	Net output summary	Description

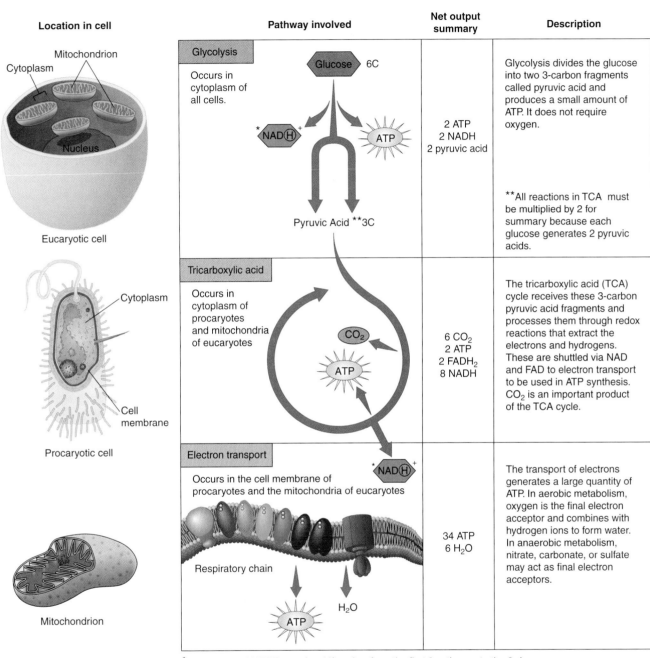

Location in cell

Mitochondrion
Cytoplasm
Nucleus

Eucaryotic cell

Cytoplasm
Cell membrane

Procaryotic cell

Mitochondrion

Pathway involved

Glycolysis

Occurs in cytoplasm of all cells.

Glucose 6C

*NAD(H)+   ATP

Pyruvic Acid **3C

Tricarboxylic acid

Occurs in cytoplasm of procaryotes and mitochondria of eucaryotes

CO_2
ATP

Electron transport

Occurs in the cell membrane of procaryotes and the mitochondria of eucaryotes

*NAD(H)+

Respiratory chain

H_2O
ATP

Net output summary

2 ATP
2 NADH
2 pyruvic acid

6 CO_2
2 ATP
2 $FADH_2$
8 NADH

34 ATP
6 H_2O

Description

Glycolysis divides the glucose into two 3-carbon fragments called pyruvic acid and produces a small amount of ATP. It does not require oxygen.

**All reactions in TCA must be multiplied by 2 for summary because each glucose generates 2 pyruvic acids.

The tricarboxylic acid (TCA) cycle receives these 3-carbon pyruvic acid fragments and processes them through redox reactions that extract the electrons and hydrogens. These are shuttled via NAD and FAD to electron transport to be used in ATP synthesis. CO_2 is an important product of the TCA cycle.

The transport of electrons generates a large quantity of ATP. In aerobic metabolism, oxygen is the final electron acceptor and combines with hydrogen ions to form water. In anaerobic metabolism, nitrate, carbonate, or sulfate may act as final electron acceptors.

*Note that the NADH+ transfers H+ and e− from the first 2 pathways to the 3rd.

Overview of the flow, location, and products of pathways in aerobic respiration
Figure 8.18

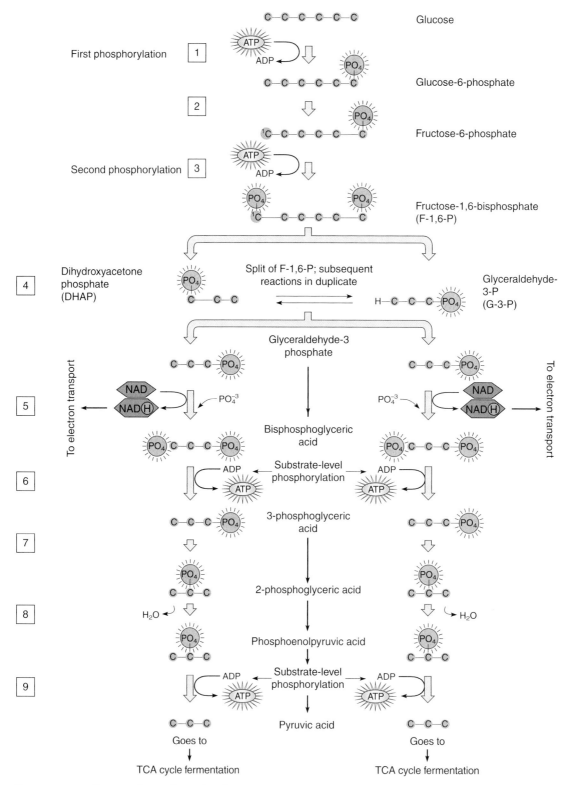

Summary figure for glycolysis
Figure 8.19

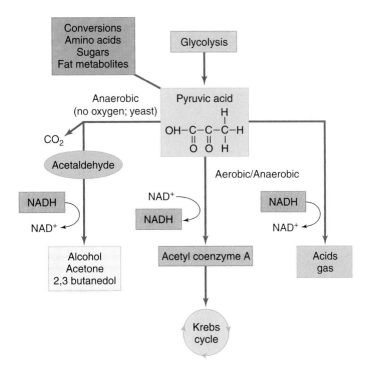

The fates of pyruvic acid (pyruvate)
Figure 8.20

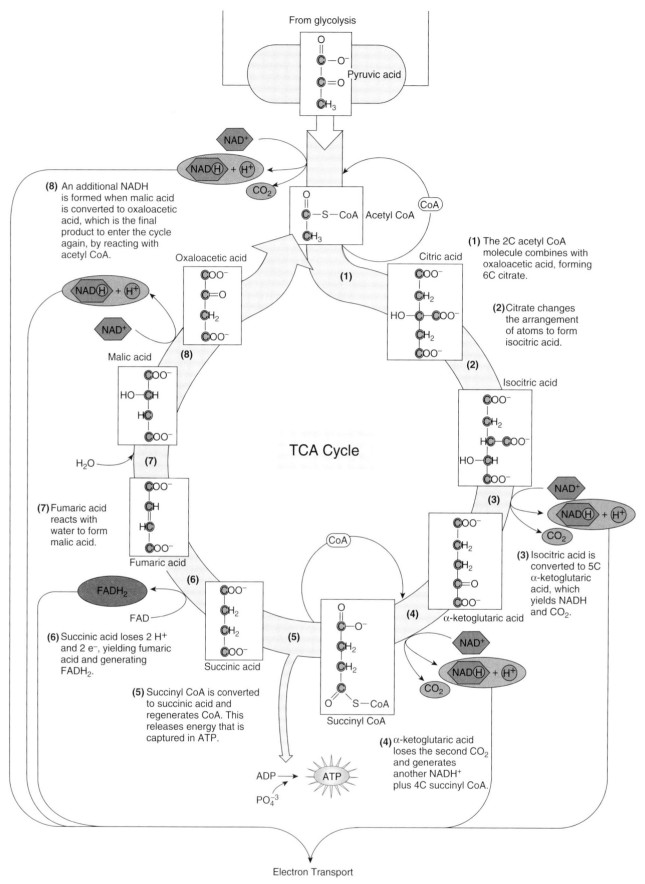

The reactions of a single turn of the TCA cycle

Figure 8.21

Adapted from Purves and Orions

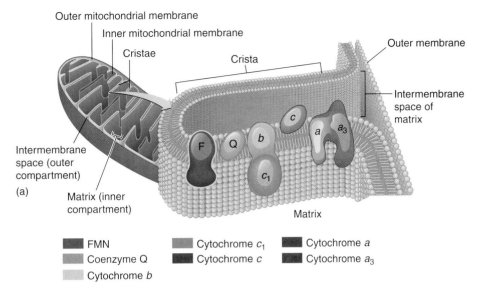

The action sites of the mitochondrion
Figure 8.22

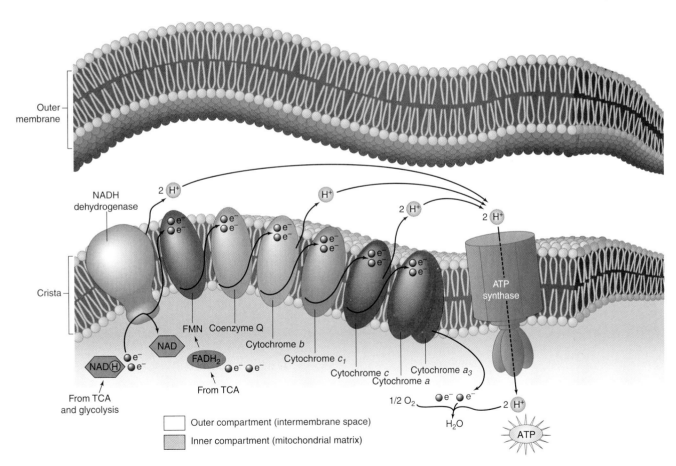

The electron transport system and oxidative phosphorylation on the mitochondrial crista
Figure 8.23

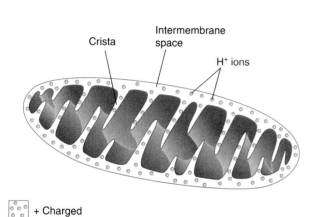

Crista

Intermembrane space

H⁺ ions

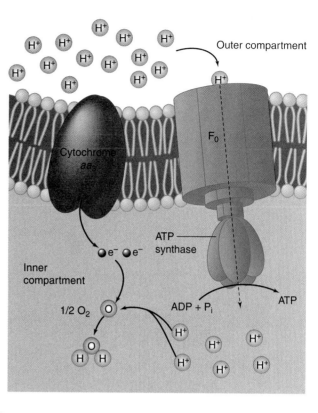

Outer compartment

H⁺

F_0

Cytochrome aa_3

ATP synthase

Inner compartment

e^- e^-

1/2 O_2

O

ADP + P_i

ATP

H

H

H⁺

H⁺

H⁺

H⁺

H⁺

H⁺

▫ + Charged

■ − Charged

(a) As the carriers in the mitochondrial cristae transport electrons, they also actively pump H⁺ ions (protons) to the intermembrane space, producing a chemical and charge gradient between the outer and inner mitochondrial compartments.

(b) The distribution of electric potential across the membrane drives the synthesis of ATP by ATP synthase. The rotation of this enzyme couples diffusion of H⁺ to the inner compartment with the bonding of ADP and P_i. The final event of electron transport is the reaction of the electrons with the accumulated H⁺ and O_2 to form metabolic H_2O. This step is catalyzed by cytochrome oxidase (cytochrome aa_3)

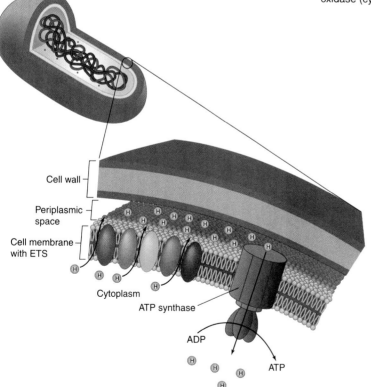

Cell wall

Periplasmic space

Cell membrane with ETS

Cytoplasm

ATP synthase

ADP

ATP

(c) Enlarged view of bacterial cell envelope to show the relationship of electron transport and ATP synthesis. Bacteria have the ETS and ATP synthase stationed in the cell membrane. ETS carriers transport H⁺ and electrons from the cytoplasm to the periplasmic space. Here, it is collected to create a gradient just as it occurs in mitochondria.

Chemiosmosis—the force behind ATP synthesis
Figure 8.24

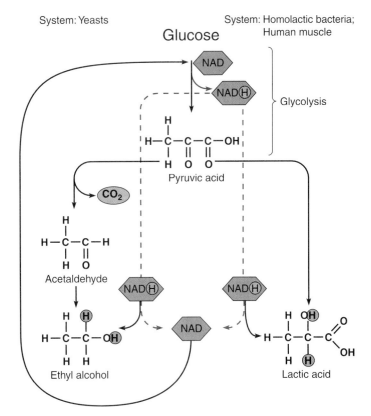

The chemistry of fermentation systems that produce acid and alcohol

Figure 8.25

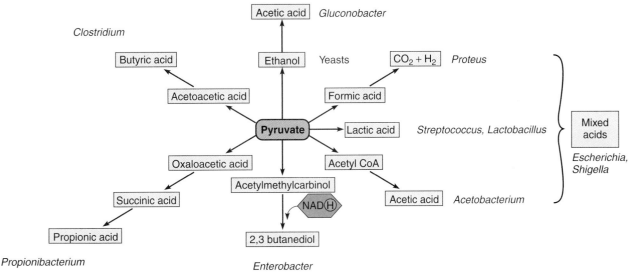

Miscellaneous products of pyruvate fermentation and the bacteria involved in their production

Figure 8.26

111

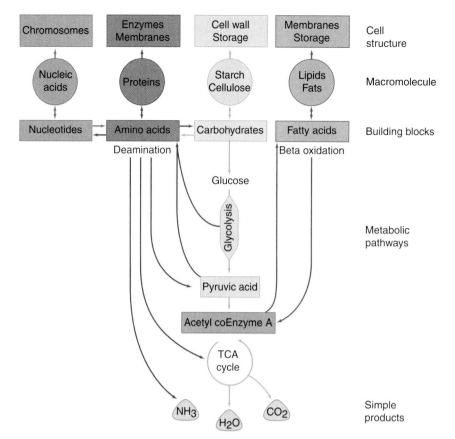

A summary of metabolic interactions
Figure 8.27

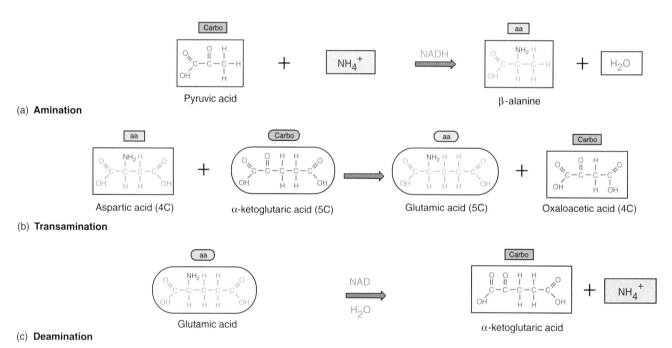

(a) **Amination**

Pyruvic acid   +   NH$_4^+$ —NADH→   β-alanine   +   H$_2$O

(b) **Transamination**

Aspartic acid (4C)   +   α-ketoglutaric acid (5C)   →   Glutamic acid (5C)   +   Oxaloacetic acid (4C)

(c) **Deamination**

Glutamic acid   —NAD, H$_2$O→   α-ketoglutaric acid   +   NH$_4^+$

Reactions that produce and convert amino acids
Figure 8.28

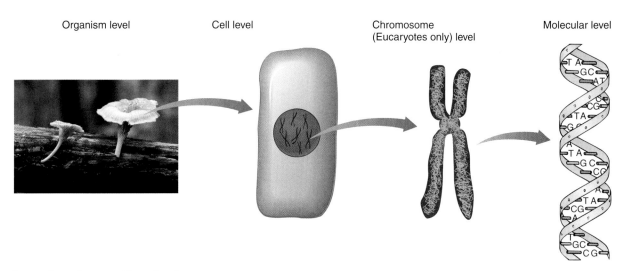

Levels of genetic study
Figure 9.1

Royalty free image from Photodisc CD VO6, Nature, Wildlife and the Environment

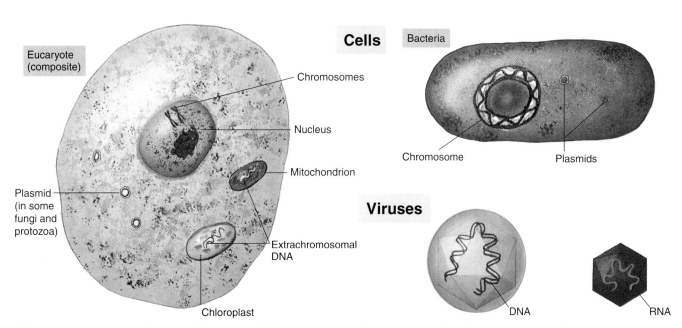

The general location and forms of the genome in selected cell types and viruses
Figure 9.2

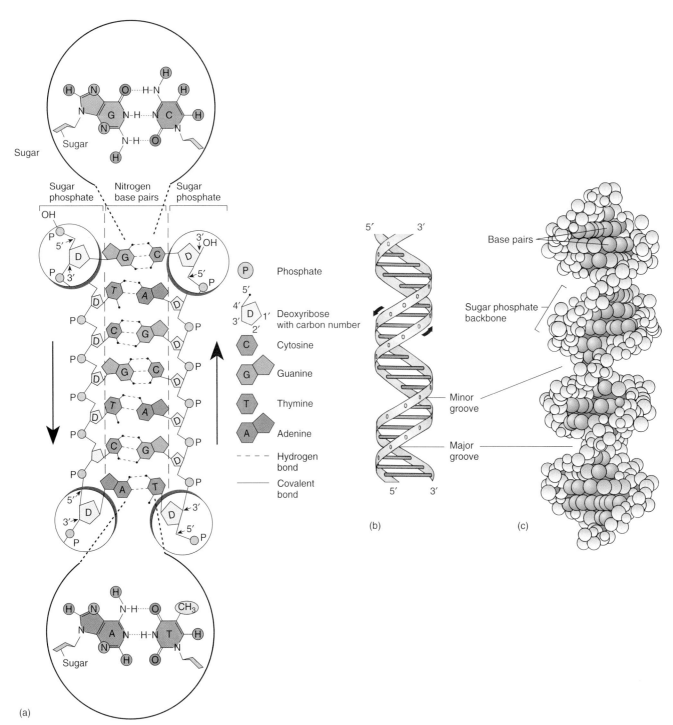

Three views of DNA structure
Figure 9.4

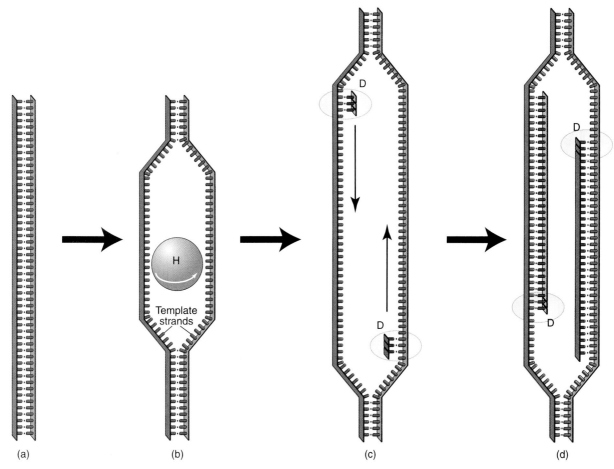

(a)   (b)   (c)   (d)

Simplified steps to show the semiconservative replication of DNA
Figure 9.5

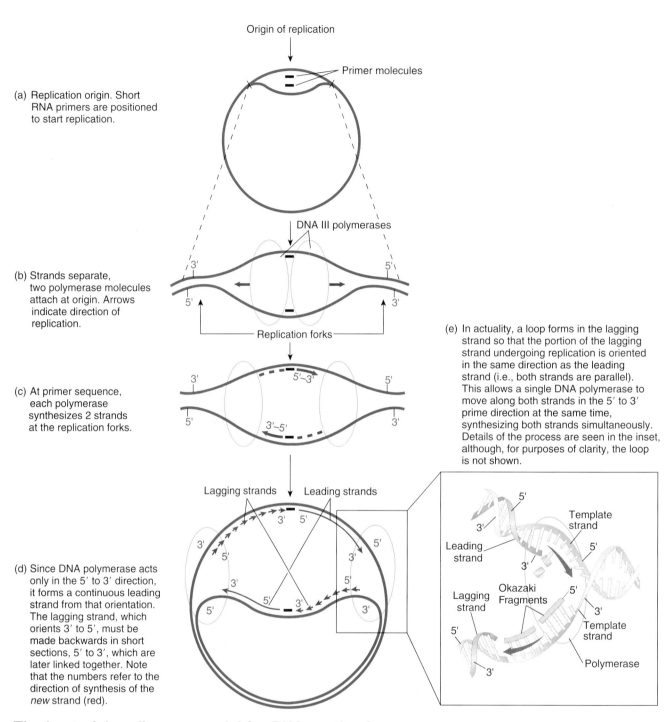

(a) Replication origin. Short RNA primers are positioned to start replication.

Origin of replication

Primer molecules

DNA III polymerases

(b) Strands separate, two polymerase molecules attach at origin. Arrows indicate direction of replication.

Replication forks

(c) At primer sequence, each polymerase synthesizes 2 strands at the replication forks.

(e) In actuality, a loop forms in the lagging strand so that the portion of the lagging strand undergoing replication is oriented in the same direction as the leading strand (i.e., both strands are parallel). This allows a single DNA polymerase to move along both strands in the 5′ to 3′ prime direction at the same time, synthesizing both strands simultaneously. Details of the process are seen in the inset, although, for purposes of clarity, the loop is not shown.

Lagging strands   Leading strands

(d) Since DNA polymerase acts only in the 5′ to 3′ direction, it forms a continuous leading strand from that orientation. The lagging strand, which orients 3′ to 5′, must be made backwards in short sections, 5′ to 3′, which are later linked together. Note that the numbers refer to the direction of synthesis of the *new* strand (red).

Template strand

Leading strand

Lagging strand

Okazaki Fragments

Template strand

Polymerase

The bacterial replicon: a model for DNA synthesis
Figure 9.6

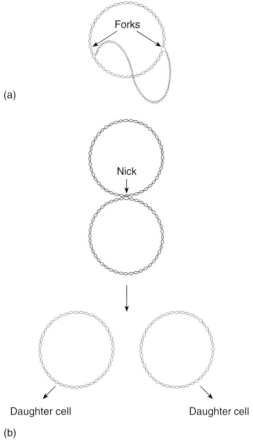

(a)

Nick

Daughter cell   Daughter cell

(b)

**Completion of chromosome
replication in bacteria**
Figure 9.7

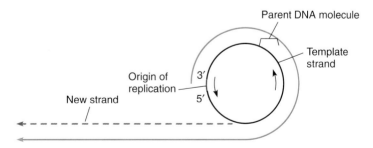

**Simplified model of rolling circle DNA
replication**
Figure 9.8

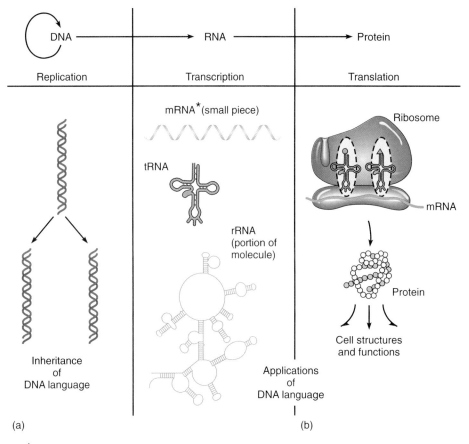

DNA → RNA → Protein

Replication | Transcription | Translation

mRNA*(small piece)

tRNA

rRNA (portion of molecule)

Ribosome

mRNA

Protein

Cell structures and functions

Inheritance of DNA language

Applications of DNA language

(a)

(b)

*The sizes of RNA are not to scale—tRNA and mRNA are enlarged to show details.

Summary of the flow of genetic information in cells
Figure 9.9

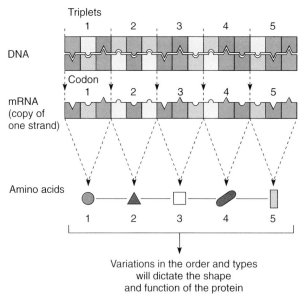

Triplets
1 2 3 4 5

DNA

Codon
1 2 3 4 5

mRNA (copy of one strand)

Amino acids
1 2 3 4 5

Variations in the order and types will dictate the shape and function of the protein

Simplified view of the DNA-protein relationship
Figure 9.10

118

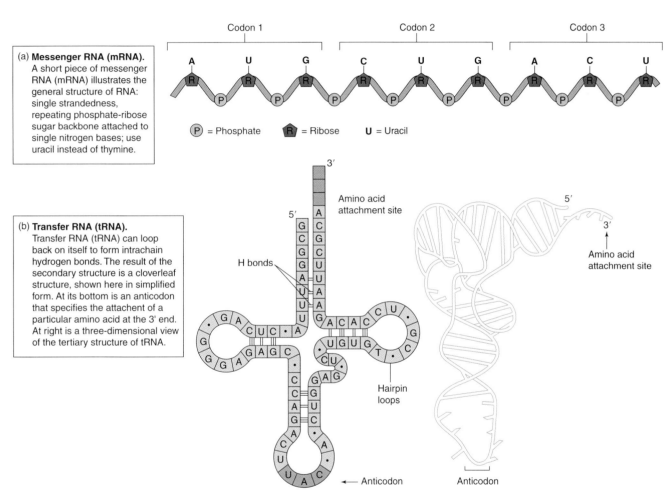

(a) **Messenger RNA (mRNA).**
A short piece of messenger RNA (mRNA) illustrates the general structure of RNA: single strandedness, repeating phosphate-ribose sugar backbone attached to single nitrogen bases; use uracil instead of thymine.

Codon 1   Codon 2   Codon 3

A U G   C U G   A C U

(P) = Phosphate   R = Ribose   U = Uracil

(b) **Transfer RNA (tRNA).**
Transfer RNA (tRNA) can loop back on itself to form intrachain hydrogen bonds. The result of the secondary structure is a cloverleaf structure, shown here in simplified form. At its bottom is an anticodon that specifies the attachent of a particular amino acid at the 3' end. At right is a three-dimensional view of the tertiary structure of tRNA.

H bonds

Amino acid attachment site

Hairpin loops

Anticodon

Amino acid attachment site

Anticodon

Characteristics of messenger and transfer RNA
Figure 9.11

119

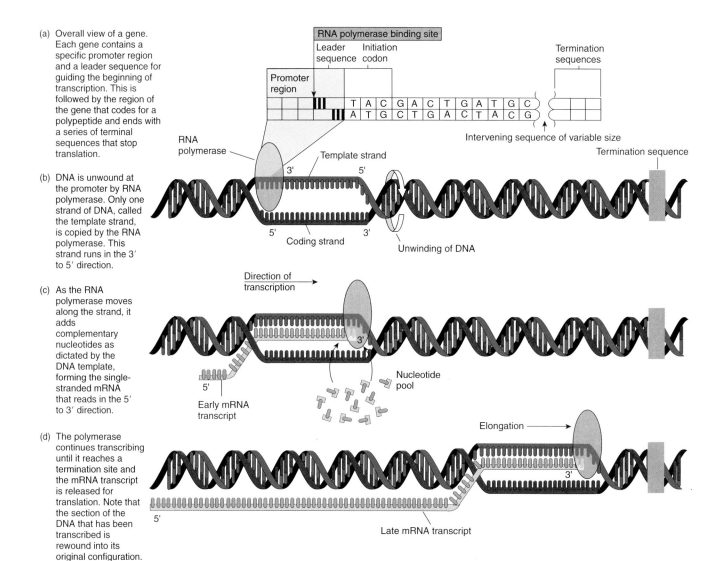

(a) Overall view of a gene. Each gene contains a specific promoter region and a leader sequence for guiding the beginning of transcription. This is followed by the region of the gene that codes for a polypeptide and ends with a series of terminal sequences that stop translation.

(b) DNA is unwound at the promoter by RNA polymerase. Only one strand of DNA, called the template strand, is copied by the RNA polymerase. This strand runs in the 3′ to 5′ direction.

(c) As the RNA polymerase moves along the strand, it adds complementary nucleotides as dictated by the DNA template, forming the single-stranded mRNA that reads in the 5′ to 3′ direction.

(d) The polymerase continues transcribing until it reaches a termination site and the mRNA transcript is released for translation. Note that the section of the DNA that has been transcribed is rewound into its original configuration.

RNA polymerase binding site

Leader sequence Initiation codon

Termination sequences

Promoter region

RNA polymerase

| T | A | C | G | A | C | T | G | A | T | G | C |
| A | T | G | C | T | G | A | C | T | A | C | G |

Intervening sequence of variable size

Template strand

Termination sequence

3′   5′

5′   3′

Coding strand

Unwinding of DNA

Direction of transcription

Nucleotide pool

5′

Early mRNA transcript

Elongation

3′

5′

Late mRNA transcript

The major events in mRNA synthesis or transcription
Figure 9.12

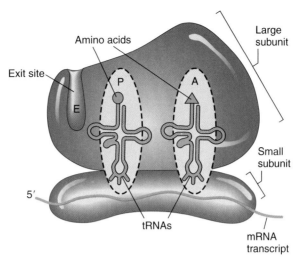

The "players" in translation
Figure 9.13

Second Position				
	U	C	A	G
U	UUU } Phenylalanine UUC UUA } Leucine UUG	UCU UCC UCA } Serine UCG	UAU } Tyrosine UAC UAA } STOP** UAG	UGU } Cysteine UGC UGA STOP** UGG Tryptophan	U C A G
C	CUU CUC CUA } Leucine CUG	CCU CCC CCA } Proline CCG	CAU } Histidine CAC CAA } Glutamine CAG	CGU CGC CGA } Arginine CGG	U C A G
A	AUU AUC } Isoleucine AUA AUG START Methionine*	ACU ACC ACA } Threonine ACG	AAU } Asparagine AAC AAA } Lysine AAG	AGU } Serine AGC AGA } Arginine AGG	U C A G
G	GUU GUC } Valine GUA GUG	GCU GCC } Alanine GCA GCG	GAU } Aspartic acid GAC GAA } Glutamic acid GAG	GGU GGC } Glycine GGA GGG	U C A G

(First Position on left; Third Position on right)

* This codon initiates translation.
**For these codons, which give the orders to stop translation, there are no corresponding tRNAs and no amino acids.

The Genetic Code: Codons of mRNA That Specify a Given Amino Acid
Figure 9.14

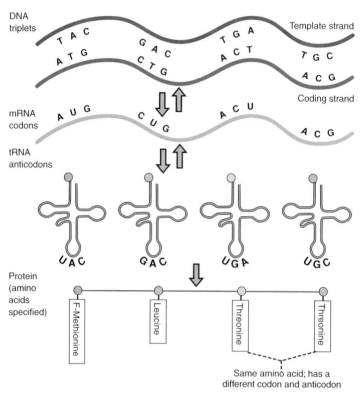

DNA triplets

Template strand

T A C   G A C   T G A

A T G   C T G   A C T   T G C

Coding strand

mRNA codons

A U G   C U G   A C U   A C G

tRNA anticodons

U A C   G A C   U G A   U G C

Protein (amino acids specified)

F-Methionine   Leucine   Threonine   Threonine

Same amino acid; has a
different codon and anticodon

Interpreting the DNA code
Figure 9.15

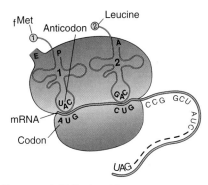

(a) Entrance of tRNAs 1 and 2

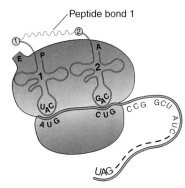

(b) Formation of peptide bond

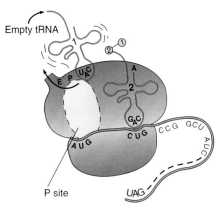

(c) Discharge of tRNA 1 at E site

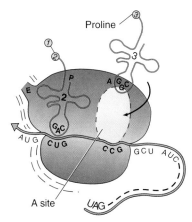

(d) First translocation; tRNA 2 shifts into P site; enter tRNA 3 by ribosome

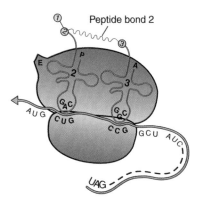

(e) Formation of peptide bond

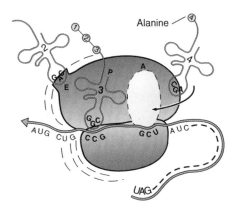

(f) Discharge of tRNA 2; second translocation; enter tRNA 4

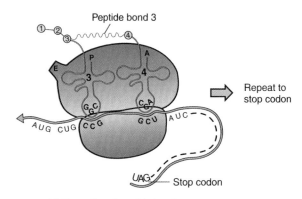

(g) Formation of peptide bond

The events in protein synthesis
Figure 9.16

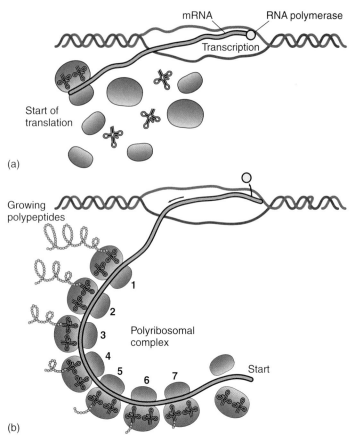

(a)

Start of translation

mRNA

RNA polymerase

Transcription

Growing polypeptides

1
2
3
4
5
6
7

Polyribosomal complex

Start

(b)

Speeding up the protein assembly line bacteria
Figure 9.17

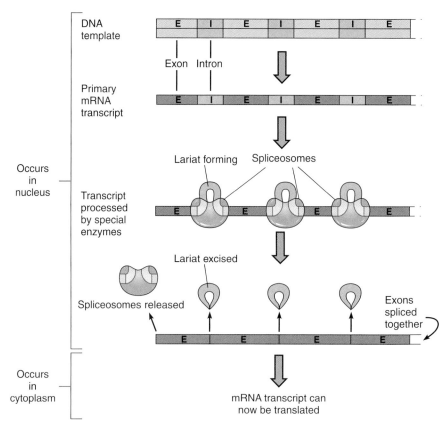

DNA template

Exon Intron

Primary mRNA transcript

Occurs in nucleus

Transcript processed by special enzymes

Lariat forming   Spliceosomes

Spliceosomes released

Lariat excised

Exons spliced together

Occurs in cytoplasm

mRNA transcript can now be translated

The split gene of eucaryotes
Figure 9.18

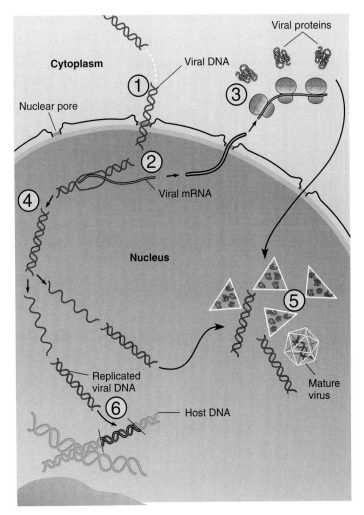

Genetic stages in the multiplication of double-stranded DNA viruses
Figure 9.19

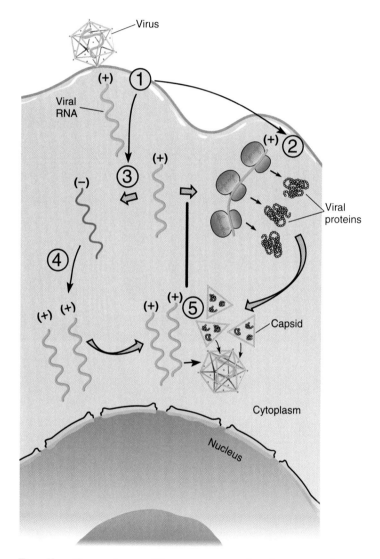

Replication of positive-sense, single-stranded RNA viruses

Figure 9.20

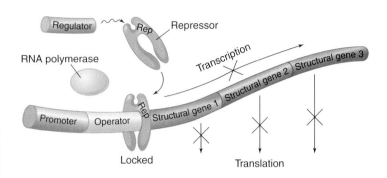

(a) Operon Off. In the absence of lactose, a repressor protein (the product of a regulatory gene located elsewhere on the bacterial chromosome) attaches to the operator gene of the operon. This effectively locks the operator and prevents any transcription of structural genes downstream (to its right). Suppression of transcription (and consequently, of translation) prevents the unnecessary synthesis of enzymes for processing lactose.

(b) Operon On. Upon entering the cell, the substrate (lactose) becomes a genetic inducer by attaching to the repressor, which loses its grip and falls away. The RNA polymerase is now free to initiate transcription, and the enzymes produced by translation of the mRNA perform the necessary reactions on their lactose substrate.

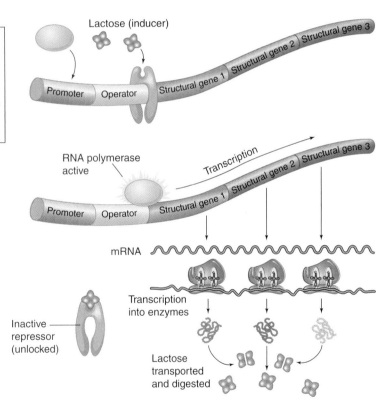

The lactose operon in bacteria: how inducible genes are controlled by substrate
Figure 9.21

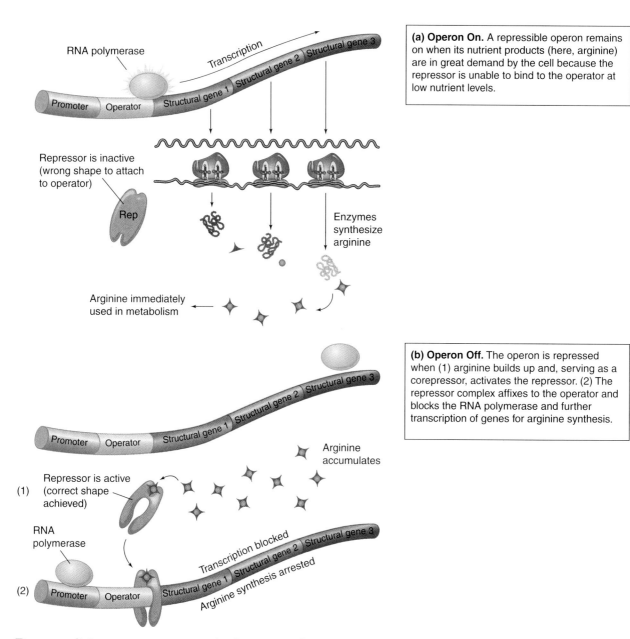

Repressible operon: control of a gene through excess nutrient
Figure 9.22

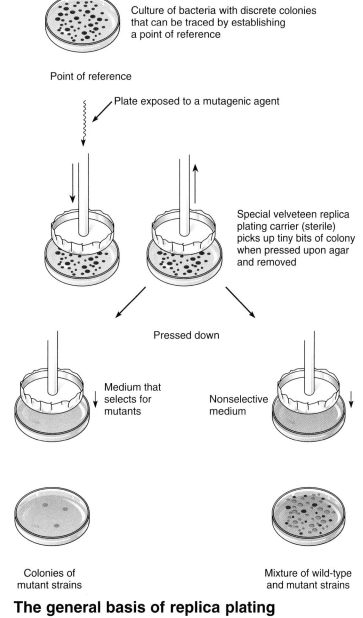

Culture of bacteria with discrete colonies that can be traced by establishing a point of reference

Point of reference

Plate exposed to a mutagenic agent

Special velveteen replica plating carrier (sterile) picks up tiny bits of colony when pressed upon agar and removed

Pressed down

Medium that selects for mutants

Nonselective medium

Colonies of mutant strains

Mixture of wild-type and mutant strains

The general basis of replica plating
Figure 9.23

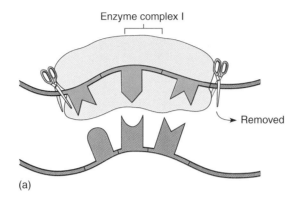

(a)

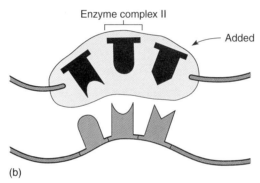

(b)

(c)

Excision repair of mutation by enzymes
Figure 9.24

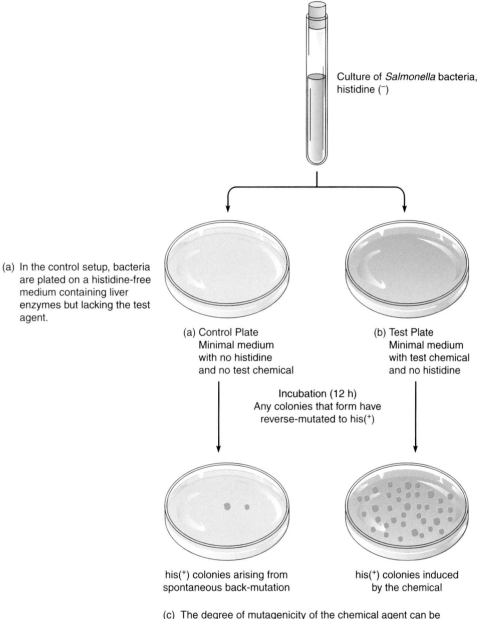

Culture of *Salmonella* bacteria, histidine (⁻)

(a) In the control setup, bacteria are plated on a histidine-free medium containing liver enzymes but lacking the test agent.

(a) Control Plate
Minimal medium
with no histidine
and no test chemical

(b) Test Plate
Minimal medium
with test chemical
and no histidine

(b) The experimental plate is prepared the same way except that it contains the test agent. After incubation, plates are observed for colonies. Any colonies developing on the plates are due to a back-mutation in a cell, which has reverted it to a his(⁺) strain.

Incubation (12 h)
Any colonies that form have
reverse-mutated to his(⁺)

his(⁺) colonies arising from
spontaneous back-mutation

his(⁺) colonies induced
by the chemical

(c) The degree of mutagenicity of the chemical agent can be calculated by comparing the number of colonies growing on the control plate with the number on the test plate. Chemicals that produce an increased incidence of back-mutation are considered carcinogens.

The Ames test
Figure 9.25

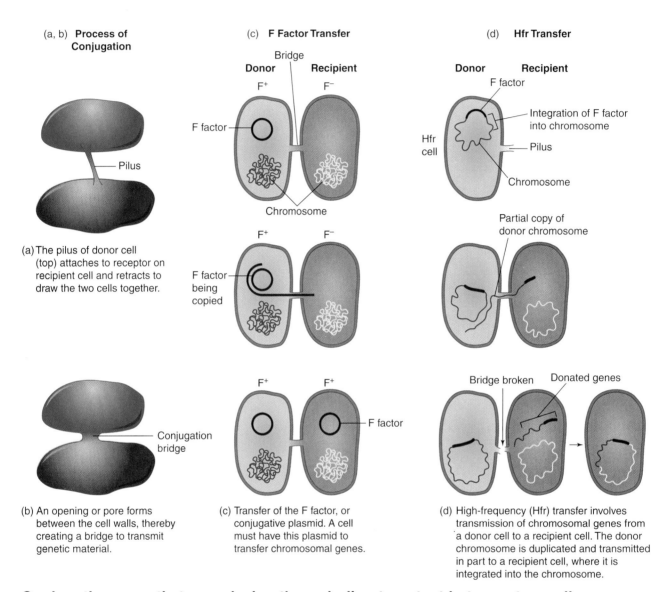

(a, b) **Process of Conjugation**	(c) **F Factor Transfer**	(d) **Hfr Transfer**

(a, b) **Process of Conjugation**

Pilus

(a) The pilus of donor cell (top) attaches to receptor on recipient cell and retracts to draw the two cells together.

Conjugation bridge

(b) An opening or pore forms between the cell walls, thereby creating a bridge to transmit genetic material.

(c) **F Factor Transfer**

Bridge

Donor   Recipient

F+   F−

F factor

Chromosome

F+   F−

F factor being copied

F+   F+

F factor

(c) Transfer of the F factor, or conjugative plasmid. A cell must have this plasmid to transfer chromosomal genes.

(d) **Hfr Transfer**

Donor   Recipient

F factor

Integration of F factor into chromosome

Hfr cell

Pilus

Chromosome

Partial copy of donor chromosome

Bridge broken   Donated genes

(d) High-frequency (Hfr) transfer involves transmission of chromosomal genes from a donor cell to a recipient cell. The donor chromosome is duplicated and transmitted in part to a recipient cell, where it is integrated into the chromosome.

Conjugation: genetic transmission through direct contact between two cells
Figure 9.26

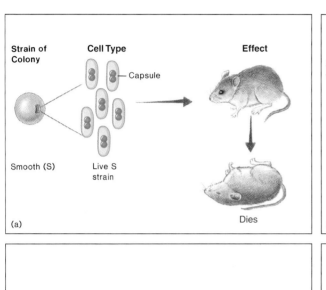

(a)

Strain of Colony

Cell Type

Effect

Capsule

Smooth (S)

Live S strain

Dies

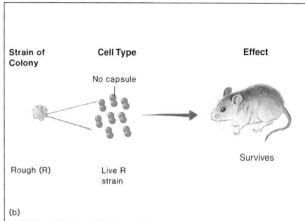

(b)

Strain of Colony

Cell Type

Effect

No capsule

Rough (R)

Live R strain

Survives

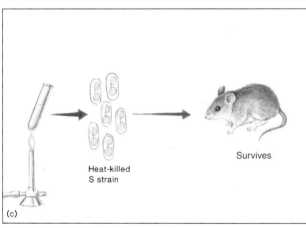

(c)

Heat-killed S strain

Survives

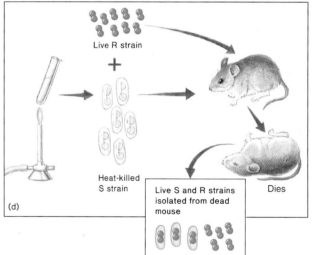

(d)

Live R strain

+

Heat-killed S strain

Live S and R strains isolated from dead mouse

Dies

Griffith's classic experiment in transformation
Figure 9.27

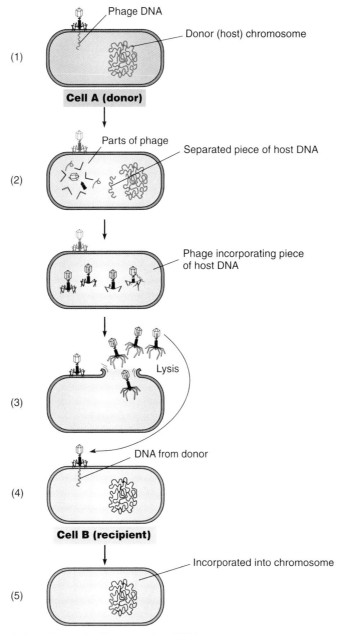

(1)

Phage DNA

Donor (host) chromosome

Cell A (donor)

(2)

Parts of phage

Separated piece of host DNA

Phage incorporating piece
of host DNA

Lysis

(3)

DNA from donor

(4)

Cell B (recipient)

Incorporated into chromosome

(5)

Cell survives and utilizes transduced DNA

**Generalized transduction: genetic transfer
by means of a virus carrier**
Figure 9.28

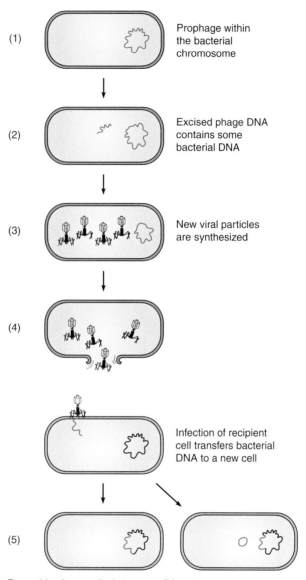

(1) Prophage within the bacterial chromosome

(2) Excised phage DNA contains some bacterial DNA

(3) New viral particles are synthesized

(4)

Infection of recipient cell transfers bacterial DNA to a new cell

(5)

Recombination results in two possible outcomes.

Specialized transduction: transfer of specific genetic material by means of a virus carrier
Figure 9.29

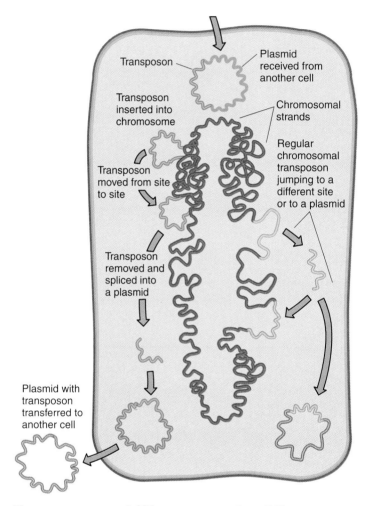

Transposon

Plasmid received from another cell

Transposon inserted into chromosome

Chromosomal strands

Regular chromosomal transposon jumping to a different site or to a plasmid

Transposon moved from site to site

Transposon removed and spliced into a plasmid

Plasmid with transposon transferred to another cell

Transposons: shifting segments of the genome
Figure 9.30

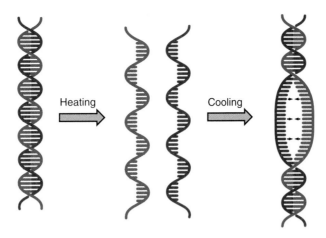

(a) **DNA heating and cooling.** DNA responds to heat by denaturing—losing its hydrogen bonding, and thereby separating into its two strands. When cooled, the two strands rejoin at complementary sites. The two strands need not be from the same organisms as long as they have matching sites.

Endonuclease	EcoRI	HindIII	HaeIII
Cutting pattern	G A A T T C C T T A A G	A A G C T T T T C G A A	G G C C C C G G

(b) **Examples of palindromes and cutting patterns.**

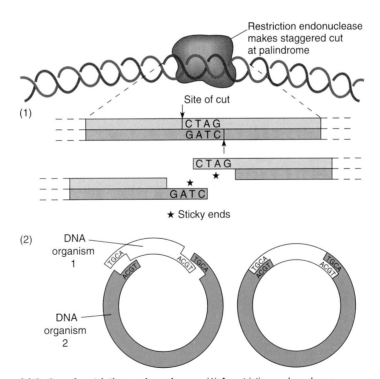

★ Sticky ends

(c) **Action of restriction endonucleases.** (1) A restriction endonuclease recognizes and cleaves DNA at the site of a specific palindromic sequence. Cleavage can produce staggered tails called sticky ends that accept complementary tails for gene splicing. (2) The sticky ends can be used to join DNA from different organisms by cutting it with the same restriction enzyme, ensuring that all fragments have complementary ends.

Some useful properties of DNA
Figure 10.1

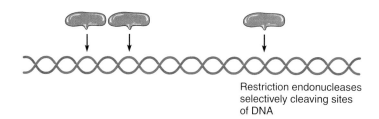

Restriction endonucleases
selectively cleaving sites
of DNA

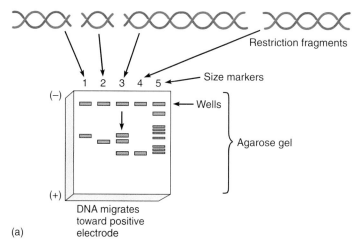

Restriction fragments

Size markers

(−)

Wells

Agarose gel

(+)

DNA migrates
toward positive
electrode

(a)

Revealing the patterns of DNA with electrophoresis
Figure 10.2

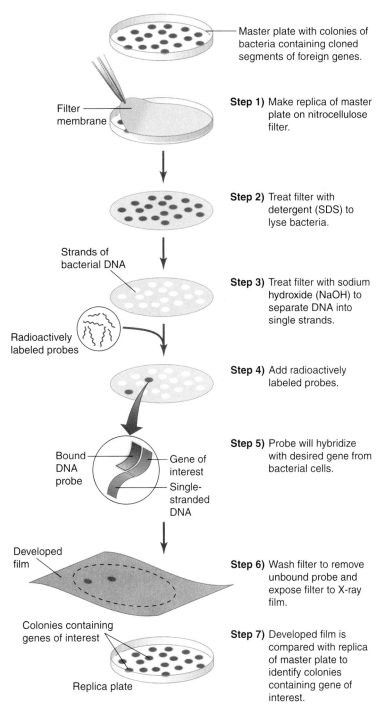

Master plate with colonies of bacteria containing cloned segments of foreign genes.

Filter membrane

Step 1) Make replica of master plate on nitrocellulose filter.

Step 2) Treat filter with detergent (SDS) to lyse bacteria.

Strands of bacterial DNA

Step 3) Treat filter with sodium hydroxide (NaOH) to separate DNA into single strands.

Radioactively labeled probes

Step 4) Add radioactively labeled probes.

Bound DNA probe

Gene of interest

Single-stranded DNA

Step 5) Probe will hybridize with desired gene from bacterial cells.

Developed film

Step 6) Wash filter to remove unbound probe and expose filter to X-ray film.

Colonies containing genes of interest

Replica plate

Step 7) Developed film is compared with replica of master plate to identify colonies containing gene of interest.

A hybridization test relies on the action of microbe-specific probes to identify an unknown bacteria or virus
Figure 10.3

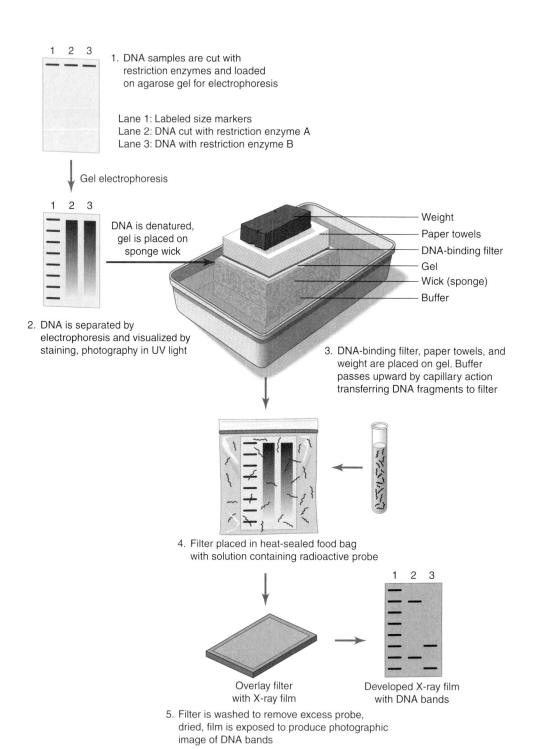

1. DNA samples are cut with restriction enzymes and loaded on agarose gel for electrophoresis

Lane 1: Labeled size markers
Lane 2: DNA cut with restriction enzyme A
Lane 3: DNA with restriction enzyme B

Gel electrophoresis

DNA is denatured, gel is placed on sponge wick

Weight
Paper towels
DNA-binding filter
Gel
Wick (sponge)
Buffer

2. DNA is separated by electrophoresis and visualized by staining, photography in UV light

3. DNA-binding filter, paper towels, and weight are placed on gel. Buffer passes upward by capillary action transferring DNA fragments to filter

4. Filter placed in heat-sealed food bag with solution containing radioactive probe

Overlay filter with X-ray film

Developed X-ray film with DNA bands

5. Filter is washed to remove excess probe, dried, film is exposed to produce photographic image of DNA bands

Conducting a Southern blot hybridization test
Figure 10.4

(1) Isolated unknown DNA fragment (one strand will be shown for clarity).

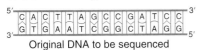

Original DNA to be sequenced

(2) DNA is denatured to produce single template strand.

(3) Strand is labeled with specific primer molecule.

(4) DNA polymerase and regular nucleotide mixture (ATP, CTP, GTP, and TTP) are added; ddG, ddA, ddC, and ddT are placed in separate reaction tubes with the regular nucleotides. The dd nucleotides are labelled with some type of tracer, which allows them to be visualized.

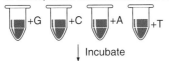

Incubate

(5) Newly replicated strands are terminated at the point of addition of a dd nucleotide.

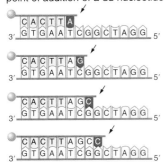

Steps in a Sanger DNA sequence technique
Figure 10.5

(6) A schematic view of how all possible positions on the fragment are occupied by a labeled nucleotide.

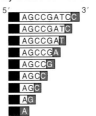

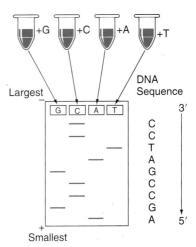

(7) Running the reaction tubes in four separate gel lanes separates them by size and nucleotide type. Reading from bottom to top, one base at a time, provides the correct DNA sequence.

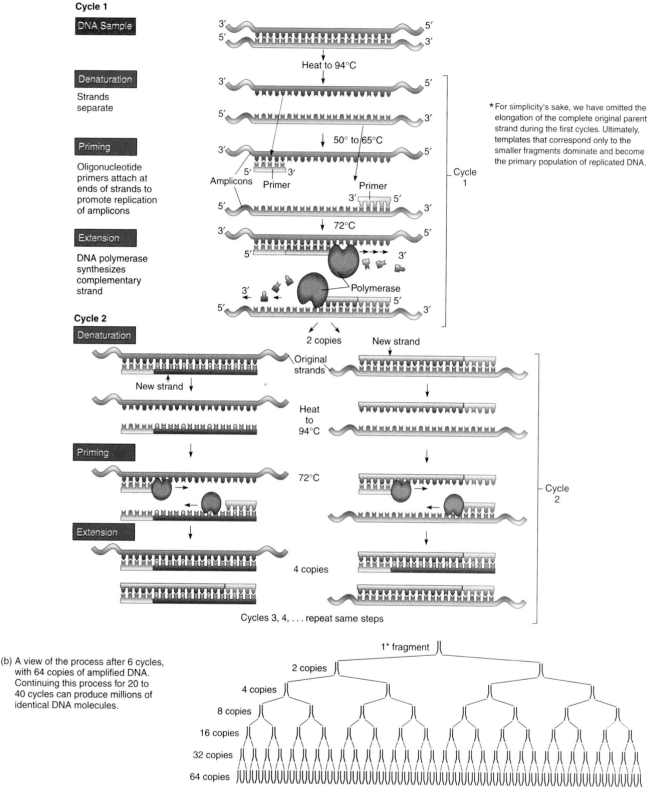

(a) In cycle 1, the DNA to be amplified is denatured, primed, and replicated by a polymerase that can function at high temperature. The two resulting strands then serve as templates for a second cycle of denaturation, priming, and synthesis.*

Cycle 1

DNA Sample

Denaturation
Strands separate

Heat to 94°C

* For simplicity's sake, we have omitted the elongation of the complete original parent strand during the first cycles. Ultimately, templates that correspond only to the smaller fragments dominate and become the primary population of replicated DNA.

50° to 65°C

Priming
Oligonucleotide primers attach at ends of strands to promote replication of amplicons

Amplicons   Primer   Primer

Cycle 1

Extension
DNA polymerase synthesizes complementary strand

72°C

Polymerase

Cycle 2

Denaturation

2 copies   New strand

Original strands

New strand

Heat to 94°C

Priming

72°C

Cycle 2

Extension

4 copies

Cycles 3, 4, . . . repeat same steps

(b) A view of the process after 6 cycles, with 64 copies of amplified DNA. Continuing this process for 20 to 40 cycles can produce millions of identical DNA molecules.

1* fragment

2 copies

4 copies

8 copies

16 copies

32 copies

64 copies

Schematic of the polymerase chain reaction
Figure 10.6

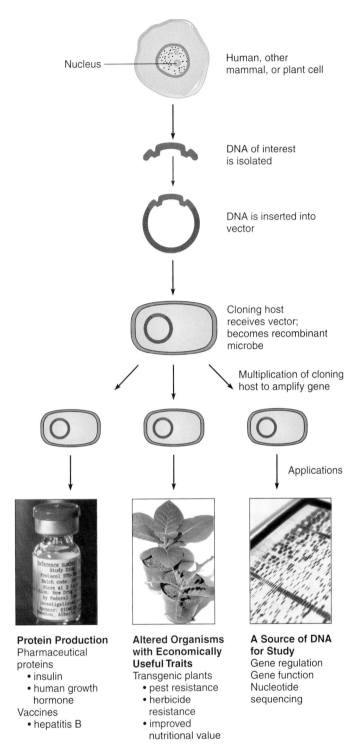

Nucleus

Human, other mammal, or plant cell

DNA of interest is isolated

DNA is inserted into vector

Cloning host receives vector; becomes recombinant microbe

Multiplication of cloning host to amplify gene

Applications

Protein Production
Pharmaceutical proteins
  • insulin
  • human growth hormone
Vaccines
  • hepatitis B

Altered Organisms with Economically Useful Traits
Transgenic plants
  • pest resistance
  • herbicide resistance
  • improved nutritional value

A Source of DNA for Study
Gene regulation
Gene function
Nucleotide sequencing

Methods and applications of genetic technology

Figure 10.7

(left): © Koester Axel/CORBIS SYGMA; (center): Milton P. Gordon, Dept. of Biochemistry, University of Washington; (right): © Andrew Brookes/CORBIS

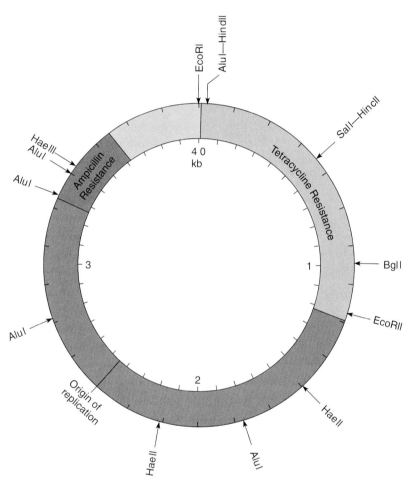

Partial map of the pBR322 plasmid of *Escherichia coli*
Figure 10.8

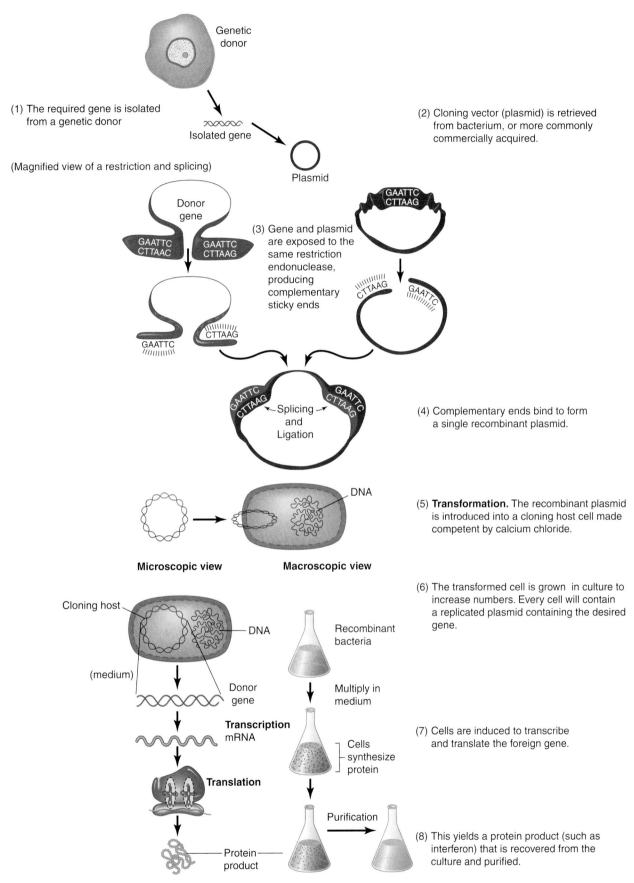

(1) The required gene is isolated from a genetic donor

Genetic donor

Isolated gene

(Magnified view of a restriction and splicing)

(2) Cloning vector (plasmid) is retrieved from bacterium, or more commonly commercially acquired.

Plasmid

Donor gene

GAATTC
CTTAAC

GAATTC
CTTAAG

GAATTC
CTTAAG

(3) Gene and plasmid are exposed to the same restriction endonuclease, producing complementary sticky ends

CTTAAG

GAATTC

GAATTC

CTTAAG

GAATTC
CTTAAG

GAATTC
CTTAAG

←Splicing→
and
Ligation

(4) Complementary ends bind to form a single recombinant plasmid.

DNA

Microscopic view   **Macroscopic view**

(5) **Transformation.** The recombinant plasmid is introduced into a cloning host cell made competent by calcium chloride.

(6) The transformed cell is grown in culture to increase numbers. Every cell will contain a replicated plasmid containing the desired gene.

Cloning host

DNA

(medium)

Donor gene

Transcription
mRNA

Translation

Protein product

Recombinant bacteria

Multiply in medium

Cells synthesize protein

Purification

(7) Cells are induced to transcribe and translate the foreign gene.

(8) This yields a protein product (such as interferon) that is recovered from the culture and purified.

Steps in recombinant DNA, gene cloning, and product retrieval
Figure 10.9

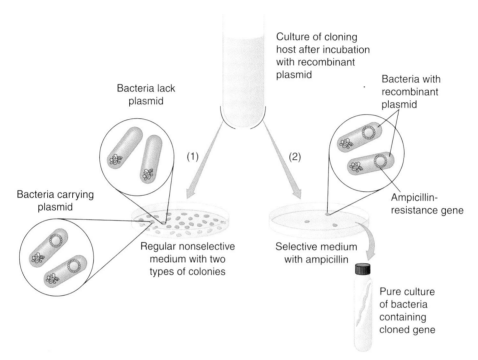

Culture of cloning host after incubation with recombinant plasmid

Bacteria lack plasmid

Bacteria with recombinant plasmid

(1)

(2)

Bacteria carrying plasmid

Ampicillin-resistance gene

Regular nonselective medium with two types of colonies

Selective medium with ampicillin

Pure culture of bacteria containing cloned gene

One method for screening clones of bacteria that have been transformed with the donor gene
Figure 10.10

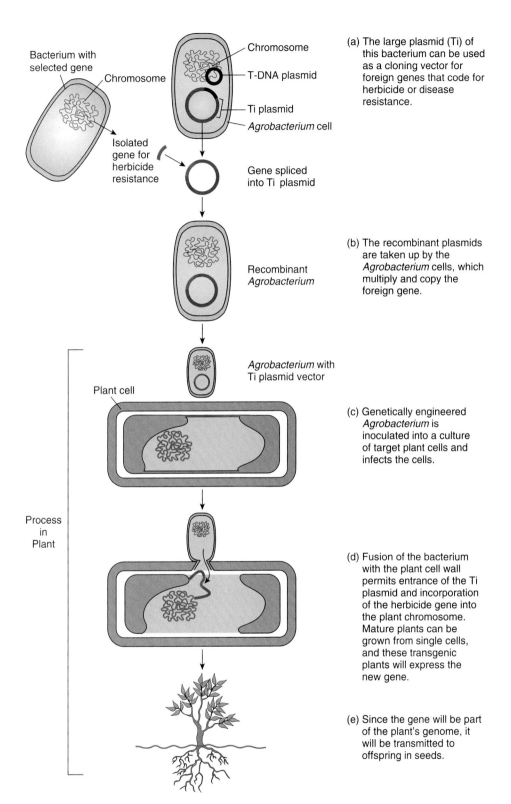

Bacterium with
selected gene

Chromosome

Chromosome

T-DNA plasmid

Ti plasmid

Agrobacterium cell

Isolated
gene for
herbicide
resistance

Gene spliced
into Ti plasmid

Recombinant
Agrobacterium

Agrobacterium with
Ti plasmid vector

Plant cell

Process
in
Plant

(a) The large plasmid (Ti) of
this bacterium can be used
as a cloning vector for
foreign genes that code for
herbicide or disease
resistance.

(b) The recombinant plasmids
are taken up by the
Agrobacterium cells, which
multiply and copy the
foreign gene.

(c) Genetically engineered
Agrobacterium is
inoculated into a culture
of target plant cells and
infects the cells.

(d) Fusion of the bacterium
with the plant cell wall
permits entrance of the Ti
plasmid and incorporation
of the herbicide gene into
the plant chromosome.
Mature plants can be
grown from single cells,
and these transgenic
plants will express the
new gene.

(e) Since the gene will be part
of the plant's genome, it
will be transmitted to
offspring in seeds.

Bioengineering of plants
Figure 10.11

Several embryos recovered from sacrificed female

Embryos transferred to a depression slide containing culture medium

Culture medium ———
Oil ———

As embryo is held in place, DNA is injected into pronucleus.

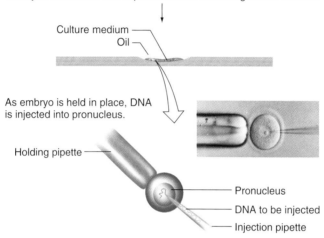

Holding pipette ———

———— Pronucleus

———— DNA to be injected

———— Injection pipette

Several injected embryos are placed into oviduct of receptive female.

How transgenic mice are created
Figure 10.12

Courtesy Brigid Hogan, Howard Hughes Medical Institute, Vanderbilt University

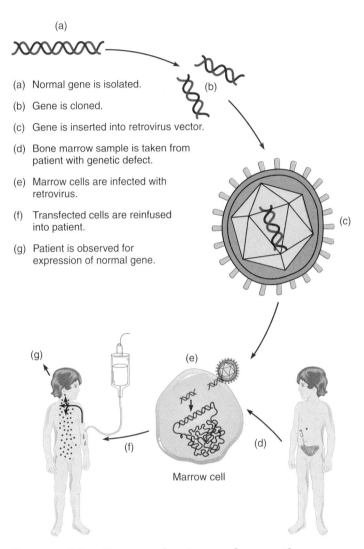

(a)

(a) Normal gene is isolated.

(b) Gene is cloned.

(c) Gene is inserted into retrovirus vector.

(d) Bone marrow sample is taken from patient with genetic defect.

(e) Marrow cells are infected with retrovirus.

(f) Transfected cells are reinfused into patient.

(g) Patient is observed for expression of normal gene.

Marrow cell

Protocol for the *ex vivo* type of gene therapy in humans

Figure 10.13

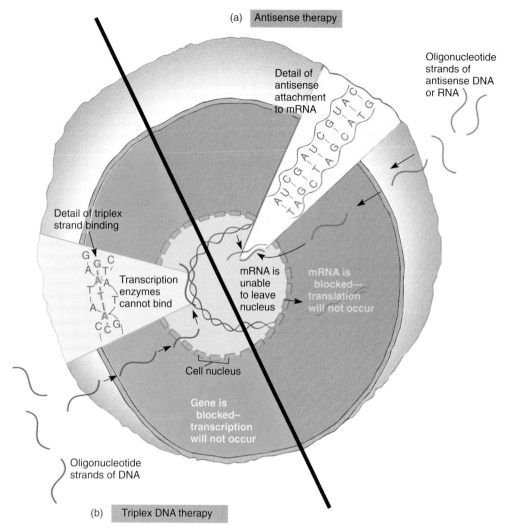

(a) Antisense therapy

Detail of
antisense
attachment
to mRNA

Oligonucleotide
strands of
antisense DNA
or RNA

Detail of triplex
strand binding

Transcription
enzymes
cannot bind

mRNA is
unable
to leave
nucleus

mRNA is
blocked—
translation
will not occur

Cell nucleus

Gene is
blocked–
transcription
will not occur

Oligonucleotide
strands of DNA

(b) Triplex DNA therapy

Mechanisms of antisense DNA and triplex DNA
Figure 10.14

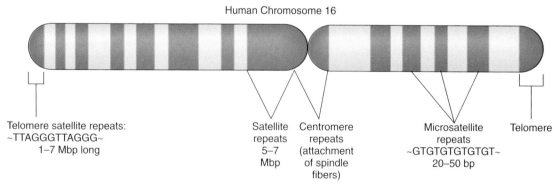

Human Chromosome 16

Telomere satellite repeats:
~TTAGGGTTAGGG~
1–7 Mbp long

Satellite
repeats
5–7
Mbp

Centromere
repeats
(attachment
of spindle
fibers)

Microsatellite
repeats
~GTGTGTGTGTGT~
20–50 bp

Telomere

A physical map of chromosome 16
Figure 10.15

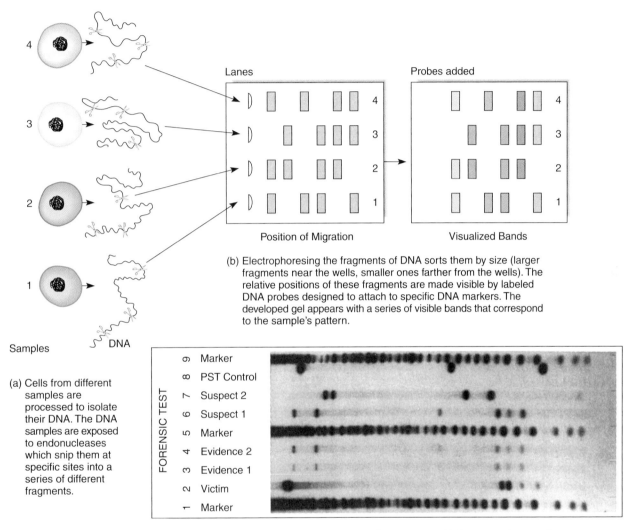

(b) Electrophoresing the fragments of DNA sorts them by size (larger fragments near the wells, smaller ones farther from the wells). The relative positions of these fragments are made visible by labeled DNA probes designed to attach to specific DNA markers. The developed gel appears with a series of visible bands that correspond to the sample's pattern.

(a) Cells from different samples are processed to isolate their DNA. The DNA samples are exposed to endonucleases which snip them at specific sites into a series of different fragments.

(c) An actual DNA fingerprint used in a rape trial. Control lanes with known markers are in lanes 1, 5, 8, and 9. The second lane contains a sample of DNA from the victim's blood. Evidence samples 1 and 2 (lanes 3 and 4) contain semen samples taken from the victim. Suspects 1 and 2 (lanes 6 and 7) were tested. Can you tell by comparing evidence and suspect lanes which individual committed the rape?

DNA fingerprints: the bar codes of life
Figure 10.16

c: Courtesy of Dr. Michael Baird, Lifecodes Corporation

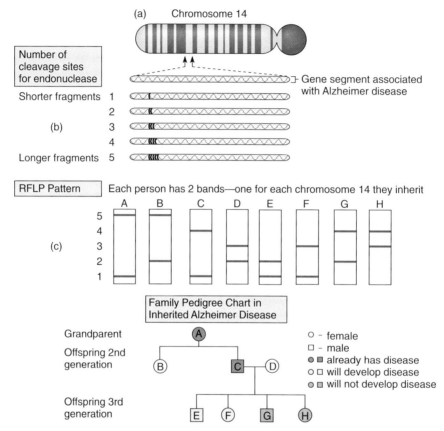

Pedigree analysis based on genetic screening for familial type Alzheimer disease
Figure 10.17

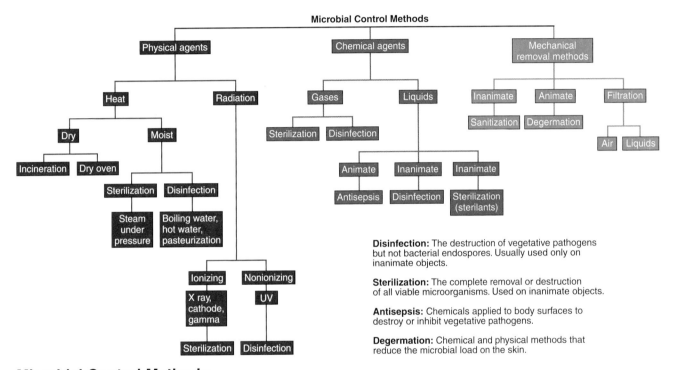

Disinfection: The destruction of vegetative pathogens but not bacterial endospores. Usually used only on inanimate objects.

Sterilization: The complete removal or destruction of all viable microorganisms. Used on inanimate objects.

Antisepsis: Chemicals applied to body surfaces to destroy or inhibit vegetative pathogens.

Degermation: Chemical and physical methods that reduce the microbial load on the skin.

Microbial Control Methods
Figure 11.1

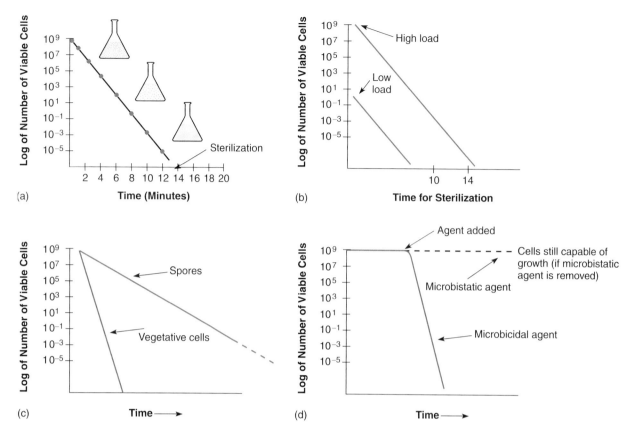

Factors that influence the rate at which microbes are killed by antimicrobial agents
Figure 11.2

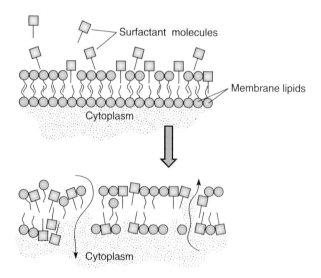

Mode of action of surfactants on the cell membrane
Figure 11.3

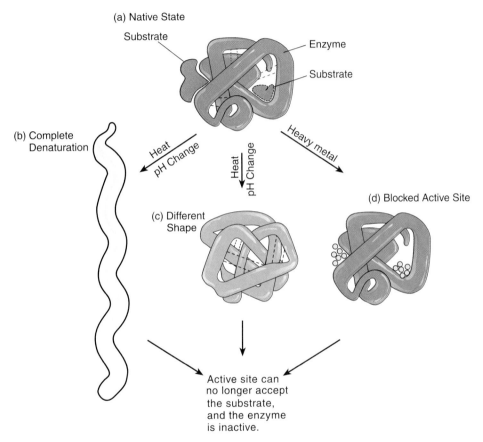

(a) Native State

Substrate

Enzyme

Substrate

(b) Complete Denaturation

Heat pH Change

Heat pH Change

Heavy metal

(d) Blocked Active Site

(c) Different Shape

Active site can no longer accept the substrate, and the enzyme is inactive.

Modes of action affecting protein function
Figure 11.4

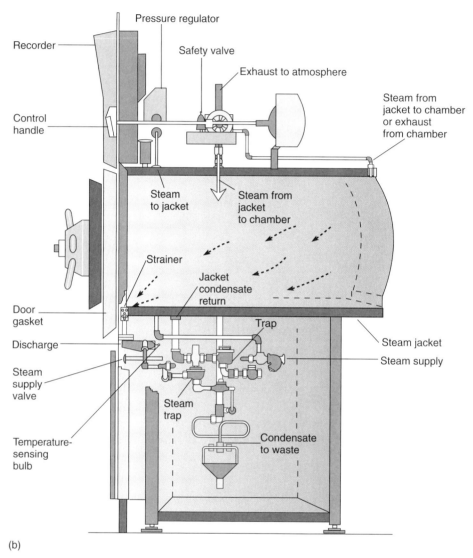

Steam sterilization with the autoclave
Figure 11.5

b: From John J. Perkins, *Principles and Methods of Sterilization in Health Science,* 2nd ed., 1969. Courtesy of Charles C. Thomas, Publisher, Springfield, Illinois.

Ionizing Radiation

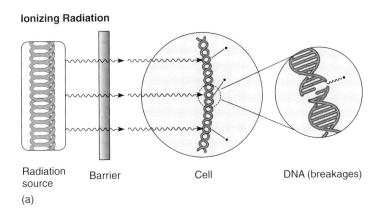

Radiation source
(a)

Barrier

Cell

DNA (breakages)

Nonionizing Radiation

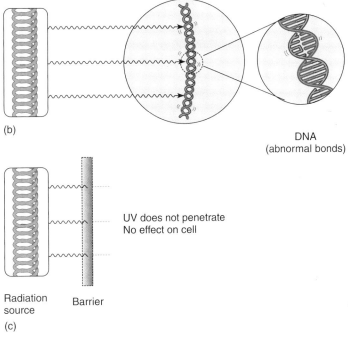

(b)

DNA
(abnormal bonds)

UV does not penetrate
No effect on cell

Radiation source
(c)

Barrier

Cellular effects of irradiation
Figure 11.7

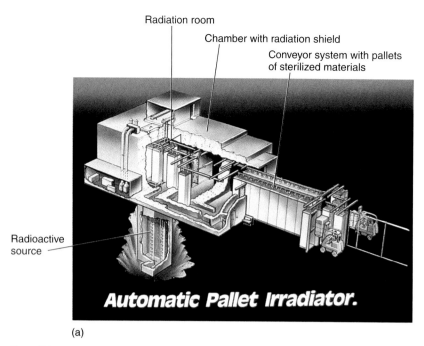

(a)

(b)

Sterilization with ionizing radiation
Figure 11.8

a: © Science VU/Nordion International/Visuals Unlimited

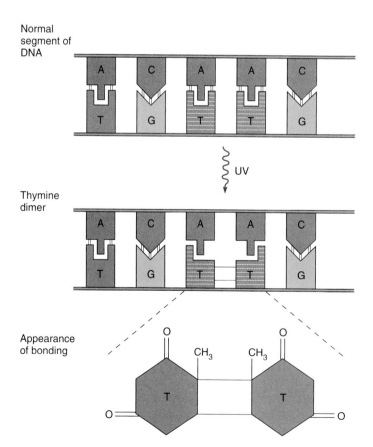

Formation of pyrimidine dimers by the action of ultraviolet (UV) radiation
Figure 11.9

An ultraviolet (UV) treatment system for disinfection of wastewater
Figure 11.10
Courtesy Trojan Technologies

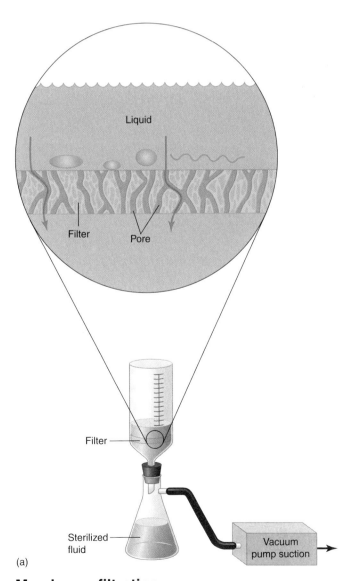

Liquid

Filter

Pore

Filter

Sterilized fluid

Vacuum pump suction

(a)

Membrane filtration
Figure 11.11

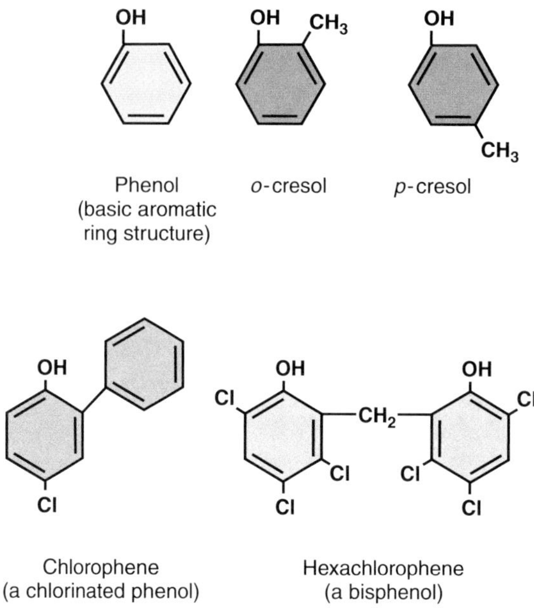

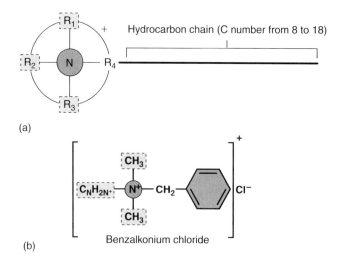

OH
OH
CH3
OH
CH3
Phenol
(basic aromatic
ring structure)
o-cresol
p-cresol
OH
Cl
OH
Cl
CH2
OH
Cl
Cl
Cl
Cl
Cl
Chlorophene
(a chlorinated phenol)
Hexachlorophene
(a bisphenol)

Some phenolics
Figure 11.12

R1
+
Hydrocarbon chain (C number from 8 to 18)
R2
N
R4
R3
(a)
+
CH3
CNH2N+
N+
CH2
Cl−
CH3
Benzalkonium chloride
(b)

The structure of detergents
Figure 11.14

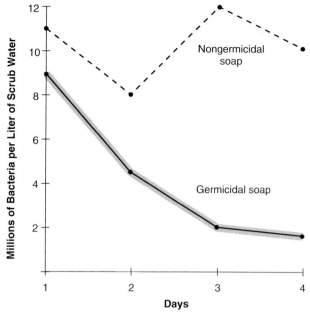

Graph showing effects of handscrubbing
Figure 11.15

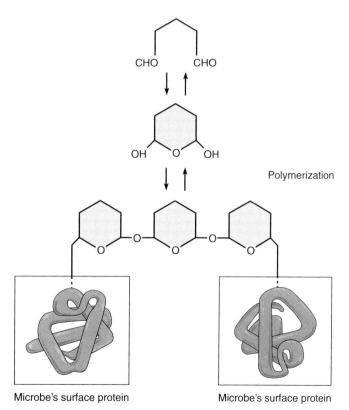

Actions of glutaraldehyde
Figure 11.17

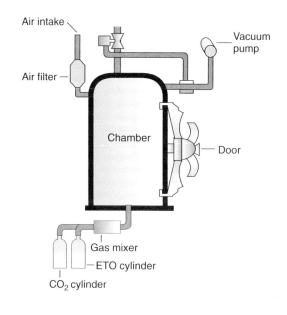

(b)

Sterilization using gas
Figure 11.18

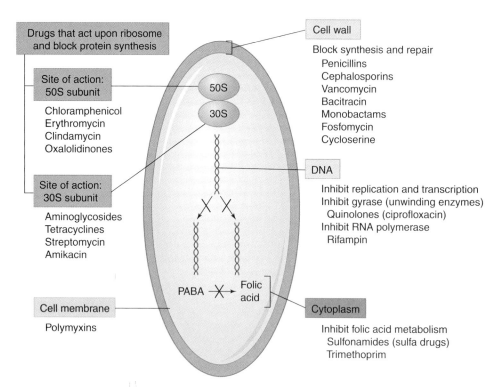

| Drugs that act upon ribosome and block protein synthesis | | Cell wall |

Cell wall

Block synthesis and repair
  Penicillins
  Cephalosporins
  Vancomycin
  Bacitracin
  Monobactams
  Fosfomycin
  Cycloserine

Site of action:
50S subunit

  Chloramphenicol
  Erythromycin
  Clindamycin
  Oxalolidinones

50S

30S

DNA

Inhibit replication and transcription
Inhibit gyrase (unwinding enzymes)
  Quinolones (ciprofloxacin)
Inhibit RNA polymerase
  Rifampin

Site of action:
30S subunit

  Aminoglycosides
  Tetracyclines
  Streptomycin
  Amikacin

PABA —X→ Folic acid

Cytoplasm

Cell membrane

  Polymyxins

Inhibit folic acid metabolism
  Sulfonamides (sulfa drugs)
  Trimethoprim

Primary sites of action of antimicrobic drugs on bacterial cells
Figure 12.1

162

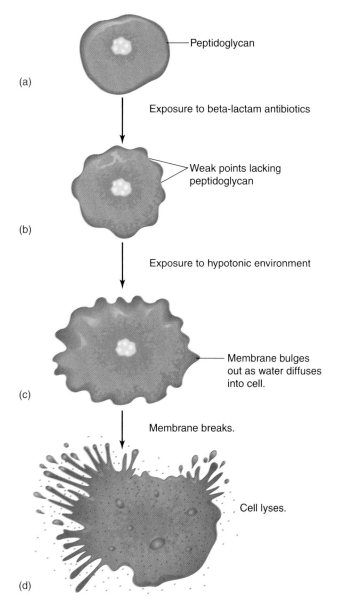

(a) Peptidoglycan

Exposure to beta-lactam antibiotics

(b) Weak points lacking peptidoglycan

Exposure to hypotonic environment

(c) Membrane bulges out as water diffuses into cell.

Membrane breaks.

(d) Cell lyses.

The consequences of exposing a growing cell to antibiotics that prevent cell wall synthesis
Figure 12.2

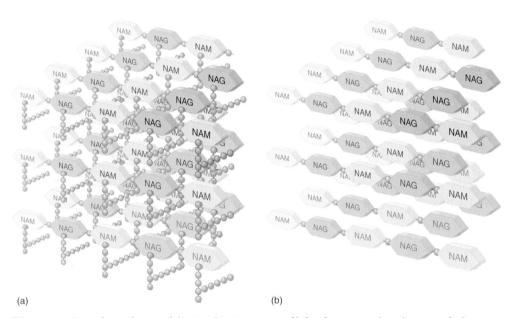

(a)

(b)

The mode of action of beta-lactam antibiotics on the bacterial cell wall

Figure 12.3

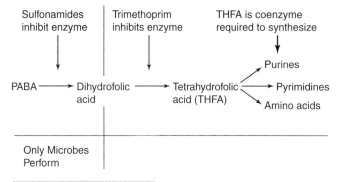

Sulfonamides inhibit enzyme

Trimethoprim inhibits enzyme

THFA is coenzyme required to synthesize

→ Purines

PABA → Dihydrofolic acid → Tetrahydrofolic acid (THFA) → Pyrimidines

→ Amino acids

Only Microbes Perform

(a) Normal metabolic pathway

COOH

NH₂
PABA

→

Active site

Pteridine synthetase

COOH

NH₂

→

COOH
O=C—NH—CH
CH₂
CH₂
COOH

NH₂
CH₂

CH

H₂N   N   N

Initial folic acid molecule

(b) Normal folic acid synthesis

COOH
NH₂
PABA

COOH
NH₂
PABA

COOH
NH₂
PABA

COOH
NH₂
PABA

COOH
NH₂
PABA

COOH
NH₂
PABA

COOH
NH₂
PABA

SO₂NH₂
NH₂
Sulfa

SO₂NH₂
NH₂
Sulfa

SO₂NH₂

Active site

NH₂ Sulfa

Pteridine synthetase

CH

N   CH₂

H₂N   N   N

Folic acid synthesis cannot be completed

(c) Inhibition of folic acid synthesis by sulfa drug

The mode of action of sulfa drugs
Figure 12.4

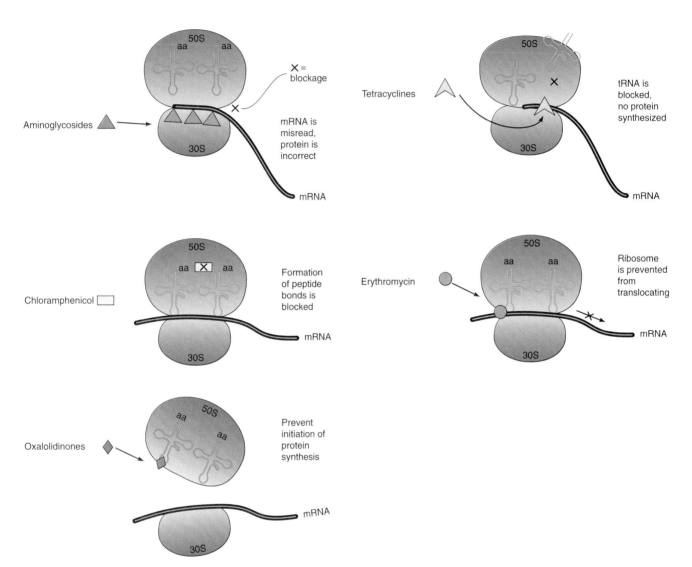

Sites of inhibition on the procaryotic ribosome and major antibiotics that act on these sites
Figure 12.5

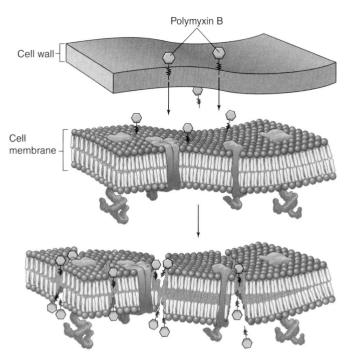

The detergent action of polymyxin
Figure 12.6

R Group

Nucleus

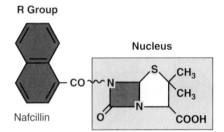

Nafcillin

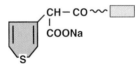

Ticarcillin

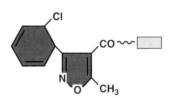

Cloxacillin

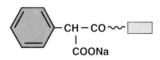

Carbenicillin

Chemical structure of penicillins
Figure 12.7

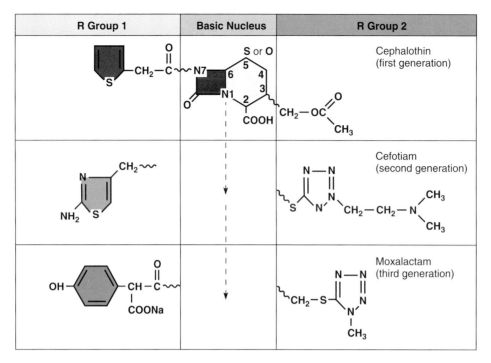

R Group 1	Basic Nucleus	R Group 2

The structure of cephalosporins

Figure 12.8

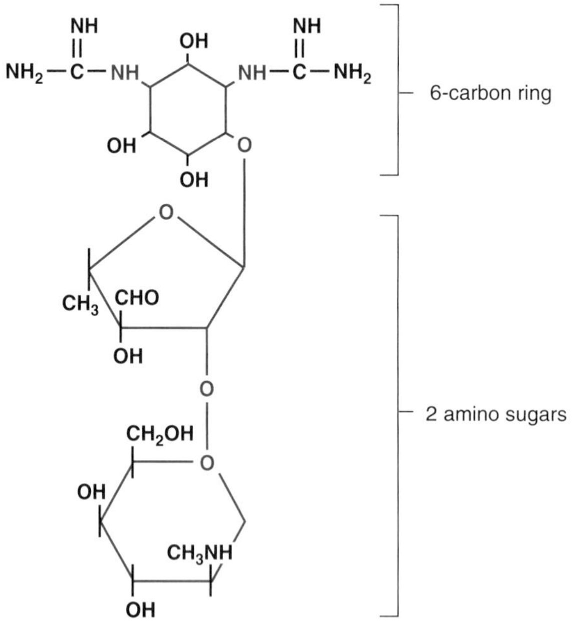

The structure of streptomycin

Figure 12.9

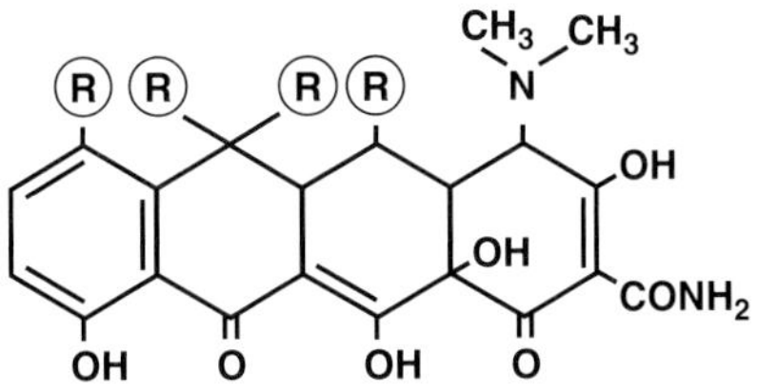

(a) **Tetracyclines**

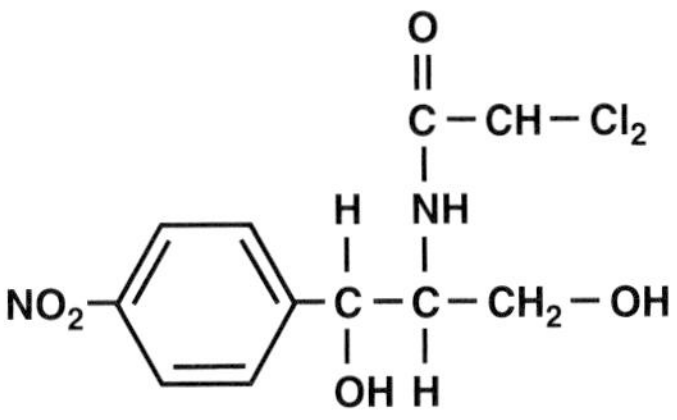

(b) **Chloramphenicol**

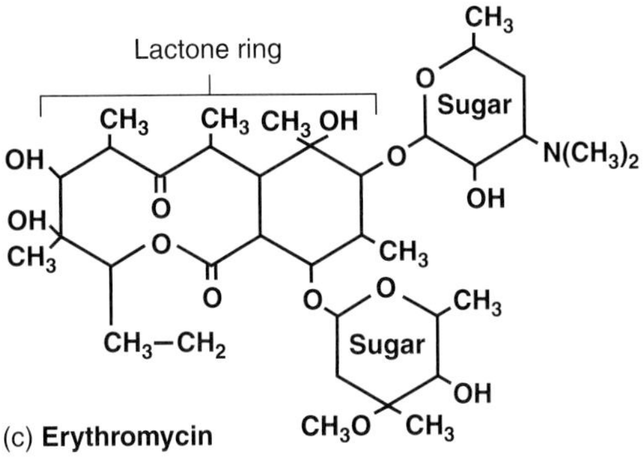
Lactone ring
Sugar
Sugar

(c) **Erythromycin**

Structures of miscellaneous broad-spectrum antibiotics
Figure 12.11

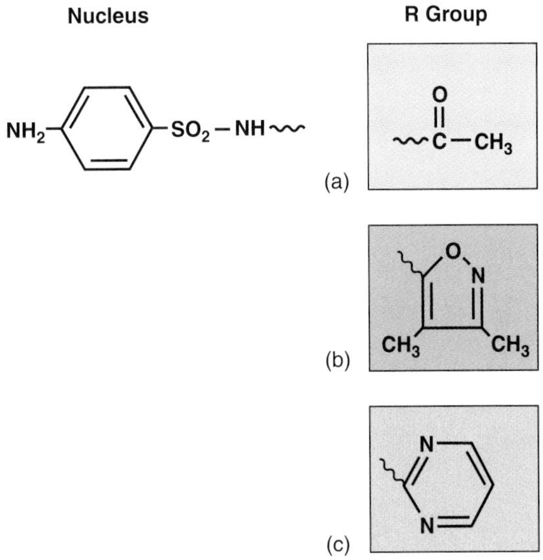
Nucleus
R Group
(a)
(b)
(c)

The structures of some sulfonamides
Figure 12.12

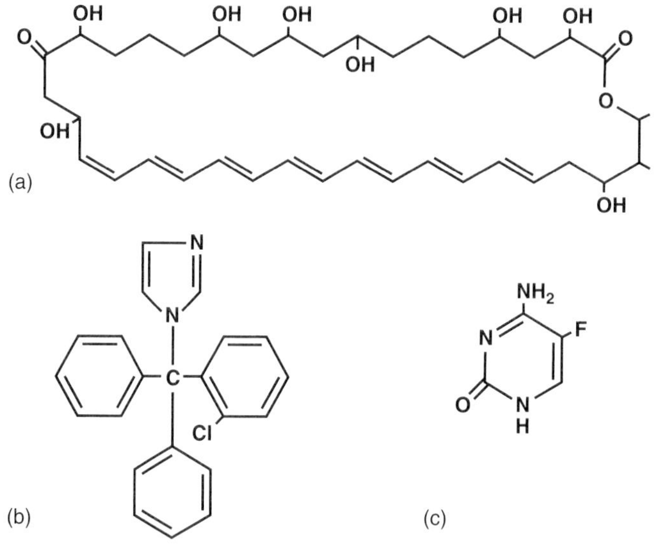

(a)

(b)

(c)

Some antifungal drug structures
Figure 12.13

TABLE 12.5

Actions of Selected Antiviral Drugs

Category	Example of Drug and Its Structure	Effects of Drug	
Inhibition of Virus Entry: receptor/fusion/uncoating inhibitors	**Fuzeon** Polypeptide of 36 amino acids (trade secret) **Amantidine** and relatives Example: Amantidine **Tamiflu, Relenza** Example: Tamiflu	(1) Blocks HIV infection by preventing the binding of viral GP-41 receptors to cell receptor, thereby preventing fusion of virus with cell (2) Block entry of influenza virus by interfering with fusion of virus with cell membrane and uncoating process (also release) Stop the actions of influenza neuraminidase, required for entry of virus into cell (also assembly)	 **No infection**
Inhibition of Nucleic Acid Synthesis	**Acyclovir,** other "cyclovirs" Example: Acyclovir **Nucleotide analog reverse transcriptase (RT) inhibitors** Example: Zidovudine (AZT) **Non-nucleoside reverse transcriptase inhibitors** Example: Nevirapine	(3) Inactivate viral DNA polymerase and terminates DNA replication in herpesviruses (4) Stop the action of reverse transcriptase in HIV, blocking viral DNA production (5) Attach to HIV RT binding site, stopping its action	 **No viral DNA synthesis** **No reverse transcription**
Inhibition of Viral Assembly/Release	**Protease inhibitors** Example: Saquinavir Amantidine and relatives; Tamiflu (see above)	(6) Insert into HIV protease, stopping its action and resulting in inactive noninfectious viruses	 **Thought to interfere with influenza virus assembly or budding**

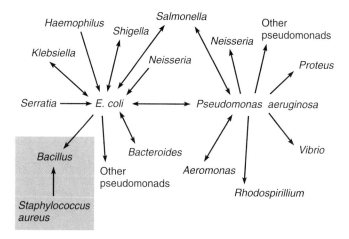

Spread of resistance factors
Figure 12.14

Source: *Data from Young and Mayer,* Review of Infectious Diseases, *1:55, 1979*

(a) Drug inactivation

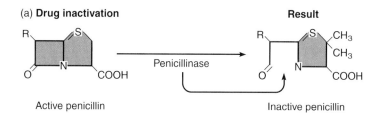

Inactivation of a drug like penicillin by penicillinase, an enzyme that cleaves a portion of the molecule and renders it inactive.

(b) Decreased permeability

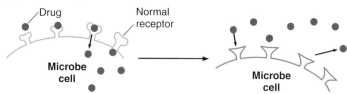

The receptor that transports the drug is altered, so that the drug cannot enter the cell.

(c) Activation of drug pumps

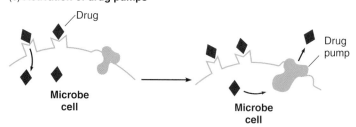

Specialized membrane proteins are activated and continually pump the drug out of the cell.

(d) Use of alternate metabolic pathway

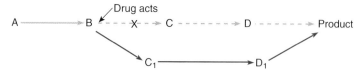

The drug has blocked the usual metabolic pathway, so the microbe circumvents it by using an alternate, unblocked pathway that achieves the required outcome.

Examples of mechanisms of acquired drug resistance
Figure 12.15

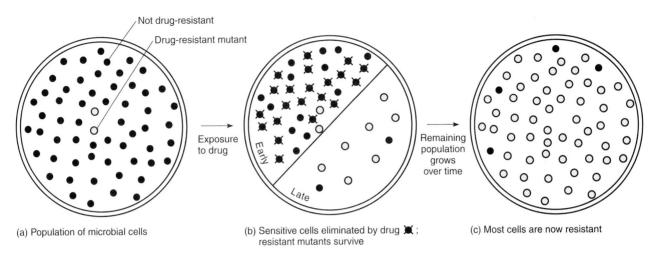

(a) Population of microbial cells

(b) Sensitive cells eliminated by drug ✖ ; resistant mutants survive

(c) Most cells are now resistant

The events in natural selection for drug resistance
Figure 12.16

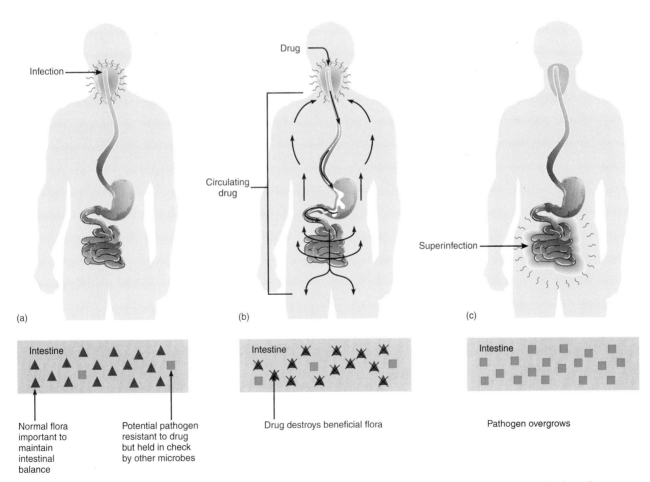

(a)

Normal flora important to maintain intestinal balance

Potential pathogen resistant to drug but held in check by other microbes

(b)

Drug destroys beneficial flora

(c)

Pathogen overgrows

The role of antimicrobics in disrupting microbial flora and causing superinfections
Figure 12.18

The Disk Diffusion Test

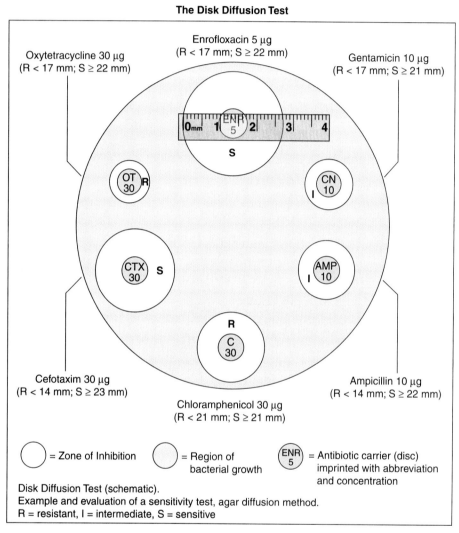

Technique for preparation and interpretation of disc diffusion tests

Figure 12.19

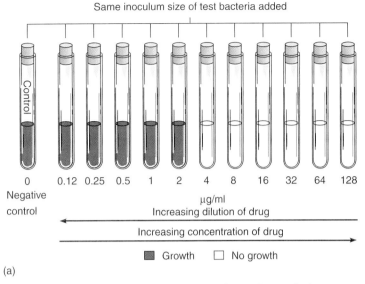

Same inoculum size of test bacteria added

Control

| 0 | 0.12 | 0.25 | 0.5 | 1 | 2 | 4 | 8 | 16 | 32 | 64 | 128 |

Negative control

µg/ml

Increasing dilution of drug

Increasing concentration of drug

■ Growth   □ No growth

(a)

Tube dilution test for determining the minimum inhibitory concentration (MIC)
Figure 12.21

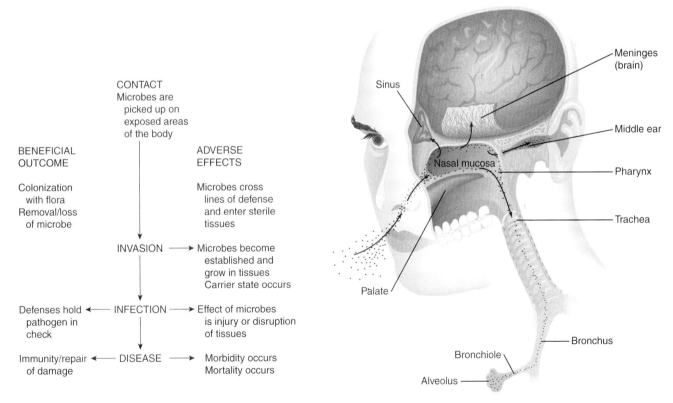

CONTACT
Microbes are picked up on exposed areas of the body

BENEFICIAL OUTCOME

Colonization with flora
Removal/loss of microbe

ADVERSE EFFECTS

Microbes cross lines of defense and enter sterile tissues

INVASION ⟶ Microbes become established and grow in tissues
Carrier state occurs

Defenses hold ← INFECTION ⟶ Effect of microbes
pathogen in check   is injury or disruption of tissues

Immunity/repair ← DISEASE ⟶ Morbidity occurs
of damage   Mortality occurs

Sinus

Meninges (brain)

Middle ear

Nasal mucosa

Pharynx

Palate

Trachea

Bronchus

Bronchiole

Alveolus

Associations between microbes and humans
Figure 13.1

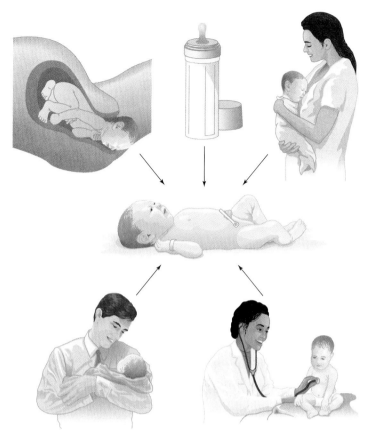

The origins of flora in newborns
Figure 13.2

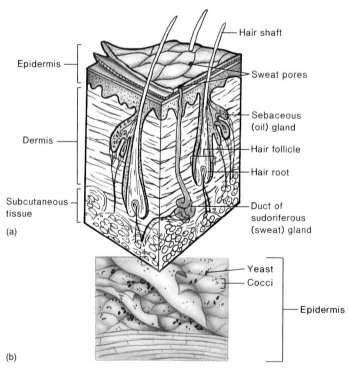

Epidermis

Dermis

Subcutaneous
tissue

(a)

Hair shaft

Sweat pores

Sebaceous
(oil) gland

Hair follicle

Hair root

Duct of
sudoriferous
(sweat) gland

Yeast
Cocci

Epidermis

(b)

The landscape of the skin
Figure 13.3

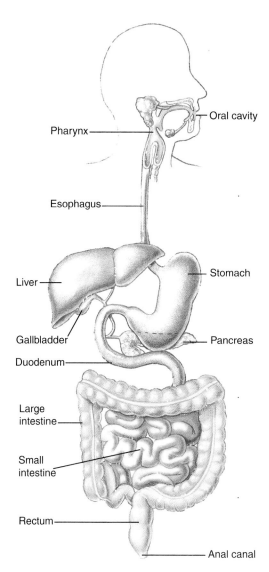

Distribution of flora
Figure 13.4

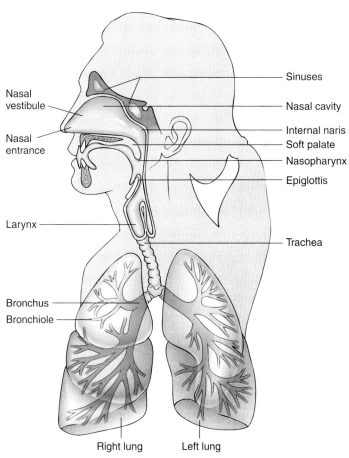

Colonized regions of the respiratory tract
Figure 13.6

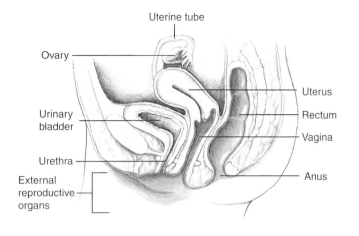

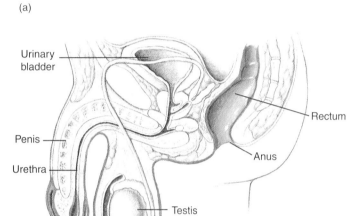

Location of (a) female and (b) male genitourinary flora (indicated by color)
Figure 13.7

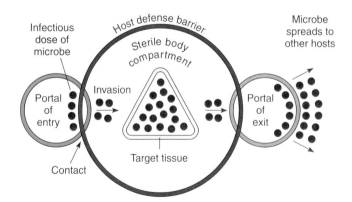

An overview of the events in infection
Figure 13.8

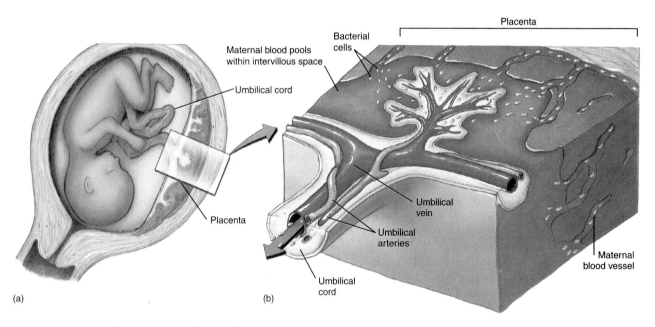

(a)

(b)

Maternal blood pools
within intervillous space

Umbilical cord

Placenta

Bacterial
cells

Placenta

Umbilical
vein

Umbilical
arteries

Umbilical
cord

Maternal
blood vessel

Transplacental infection of the fetus
Figure 13.9

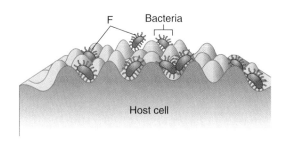

(a) **Fimbriae**

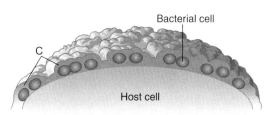

(b) **Capsules**

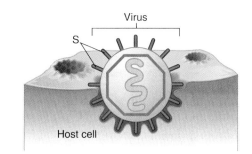

(c) **Spikes**

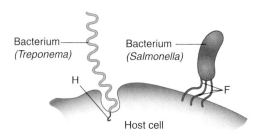

(d) **Hooks or flagella**

Mechanisms of adhesion by pathogens
Figure 13.10

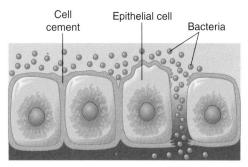

(a) **Exoenzymes**

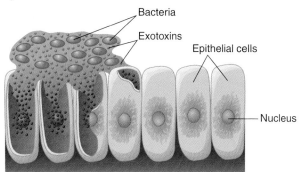

(b) **Toxins**

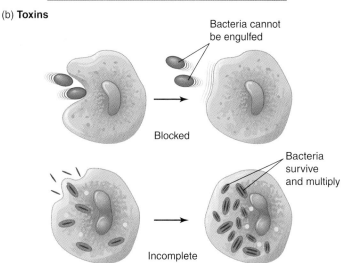

(c) **Phagocytosis**

The function of exoenzymes, toxins, and phagocyte blockers in invasiveness
Figure 13.11

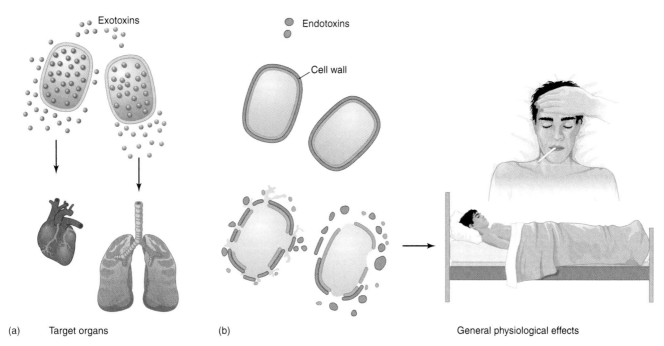

(a)   Target organs   (b)   General physiological effects

The origins and effects of circulating exotoxins and endotoxins
Figure 13.12

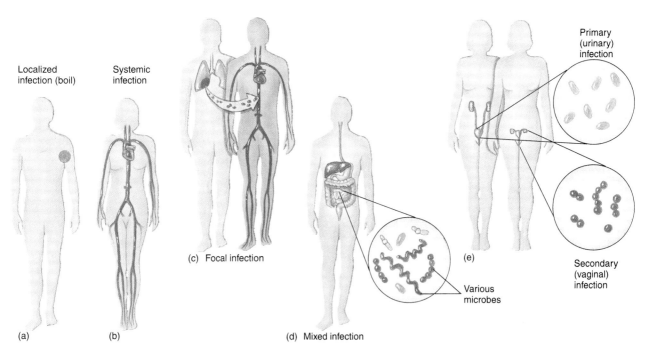

The occurrence of infections with regard to location, type of microbe, and length of time
Figure 13.13

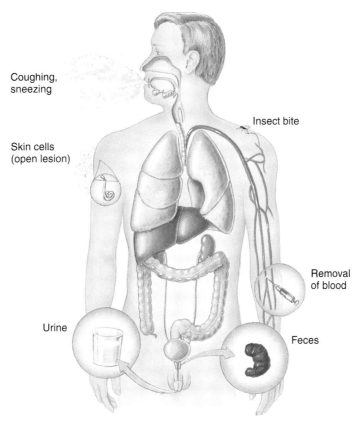

Coughing, sneezing

Skin cells (open lesion)

Insect bite

Removal of blood

Urine

Feces

Major portals of exit of infectious diseases
Figure 13.14

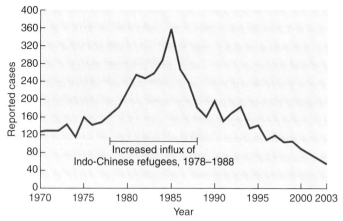

(a) **Leprosy — Reported cases by year, United States, 1970–2003**

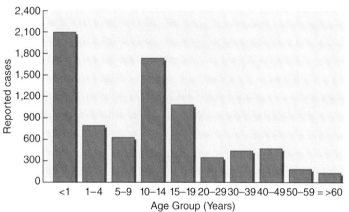

(b) **Pertussis — Reported cases by age group, United States, 2000**

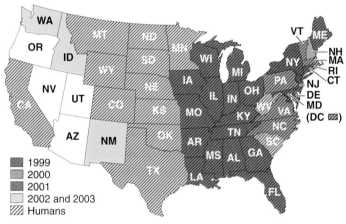

(c) West Nile virus cases were first reported in 1999 among a variety of animals and humans. Colors track its rapid progress across the United States in just four years.

Graphical representation of epidemiological data
Figure 13.15

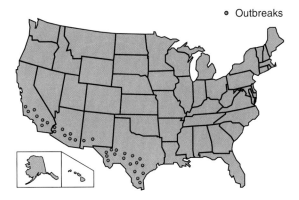

(a) **Endemic Occurrence**

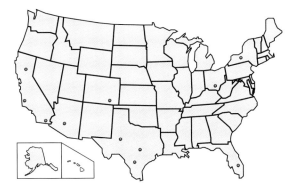

(b) **Sporadic Occurrence**

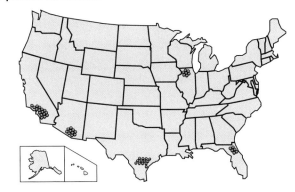

(c) **Epidemic Occurrence**

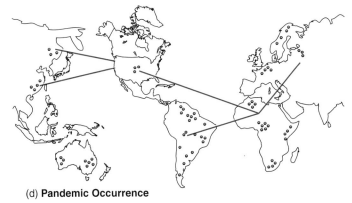

(d) **Pandemic Occurrence**

Patterns of infectious disease occurrence
Figure 13.16

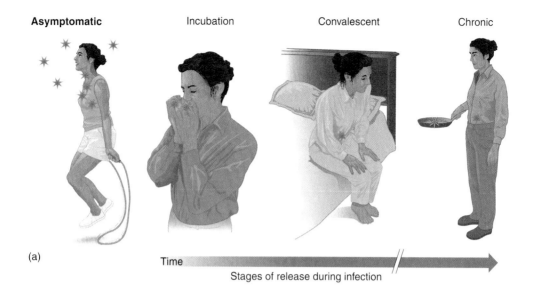

Asymptomatic   Incubation   Convalescent   Chronic

(a)

Time

Stages of release during infection

Passive

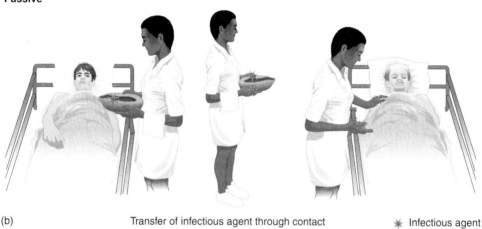

(b)   Transfer of infectious agent through contact   ✳ Infectious agent

Types of carriers
Figure 13.17

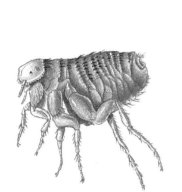

(a) Biological vectors are infected.

(b) Mechanical vectors are not infected.

Two types of vectors (magnified views)
Figure 13.18

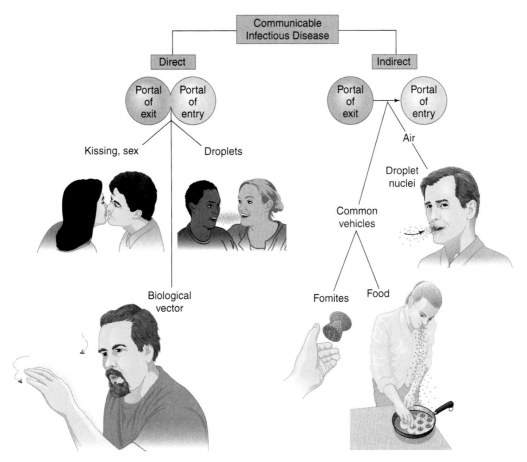

Summary of how communicable infectious diseases are acquired
Figure 13.19

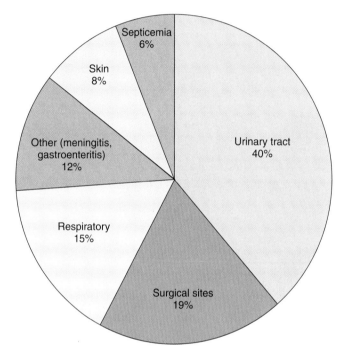

Most common nosocomial infections
Figure 13.21

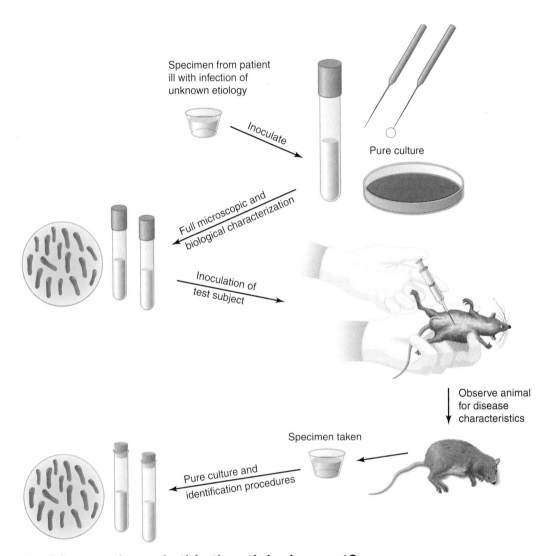

Specimen from patient ill with infection of unknown etiology

Inoculate

Pure culture

Full microscopic and biological characterization

Inoculation of test subject

Observe animal for disease characteristics

Specimen taken

Pure culture and identification procedures

Koch's postulates: Is this the etiologic agent?
Figure 13.22

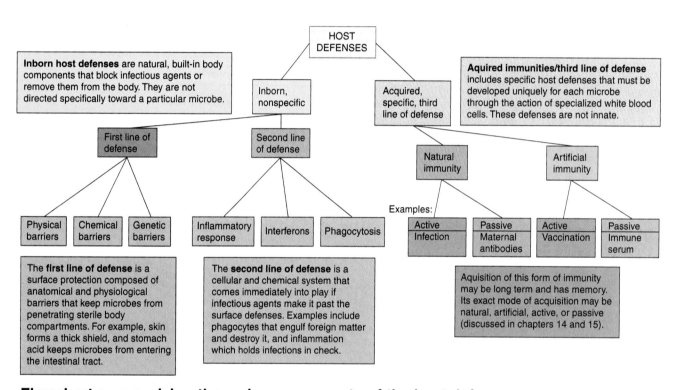

Flowchart summarizing the major components of the host defenses
Figure 14.1

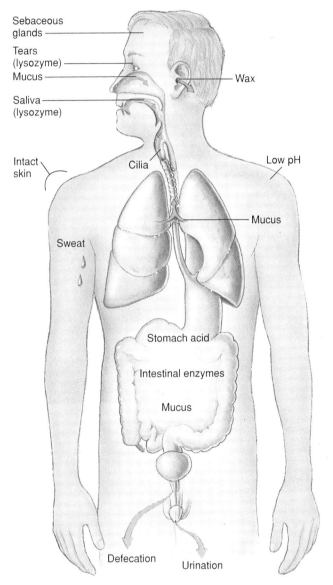

The primary physical and chemical defense barriers
Figure 14.2

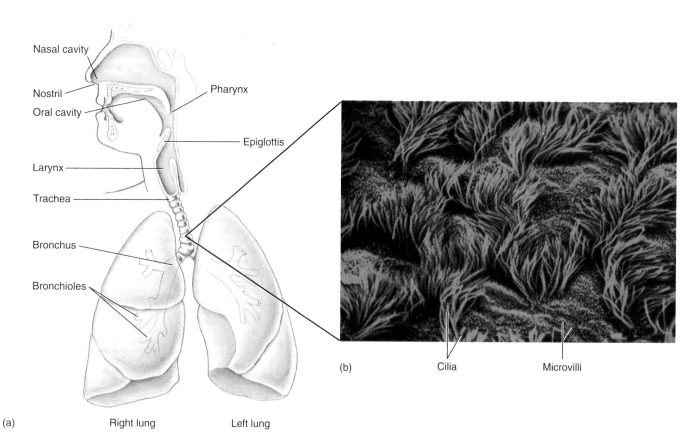

Nasal cavity

Nostril

Oral cavity

Pharynx

Epiglottis

Larynx

Trachea

Bronchus

Bronchioles

(a)

Right lung

Left lung

(b)

Cilia

Microvilli

The ciliary defense of the respiratory tree
Figure 14.3

b: © Ellen R. Dirksen/Visuals Unlimited

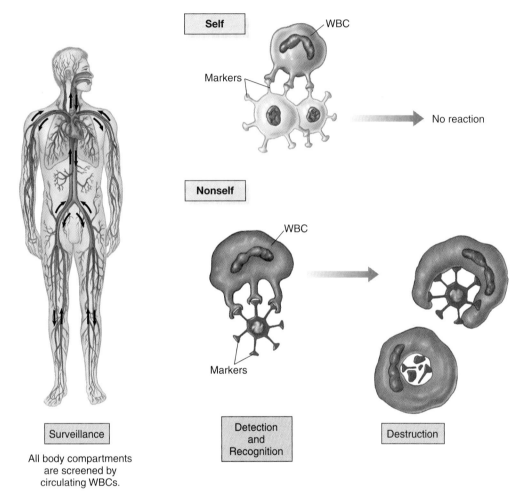

Self

WBC

Markers

No reaction

Nonself

WBC

Markers

Surveillance

All body compartments
are screened by
circulating WBCs.

Detection
and
Recognition

Destruction

**Search, recognize, and destroy is the mandate of the immune
system**
Figure 14.4

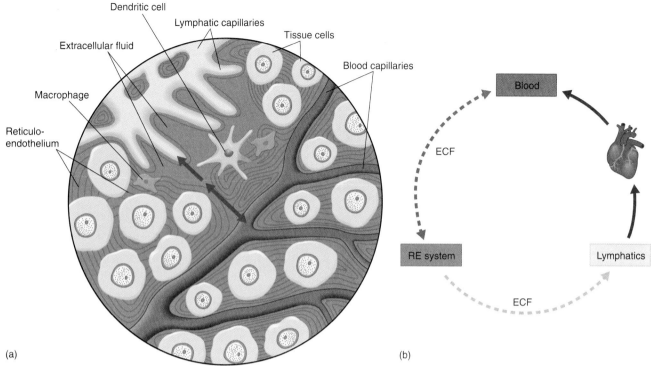

(a)

(b)

Connections between the body compartments
Figure 14.5

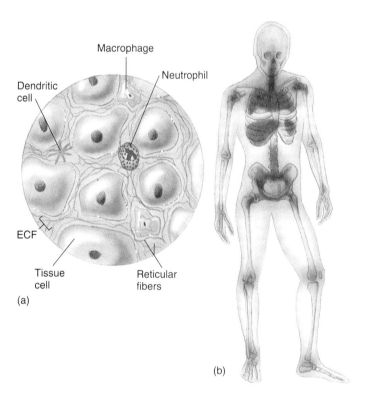

(a)

(b)

The reticuloendothelial system occurs as a pervasive, continuous connective tissue framework throughout the body
Figure 14.6

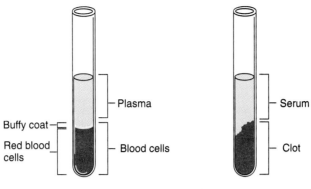

(a) **Unclotted Whole Blood**   (b) **Clotted Whole Blood**

The macroscopic composition of whole blood
Figure 14.7

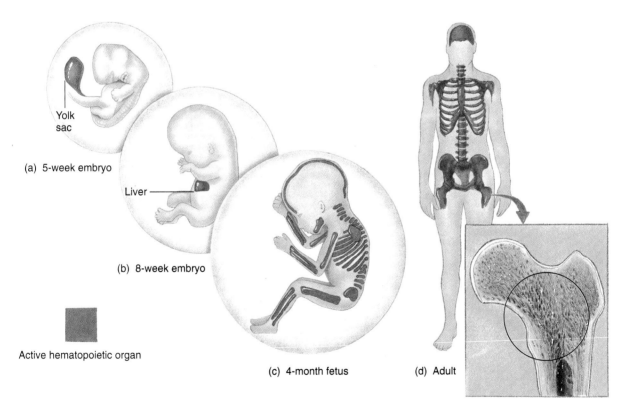

Stages in hemopoiesis
Figure 14.8

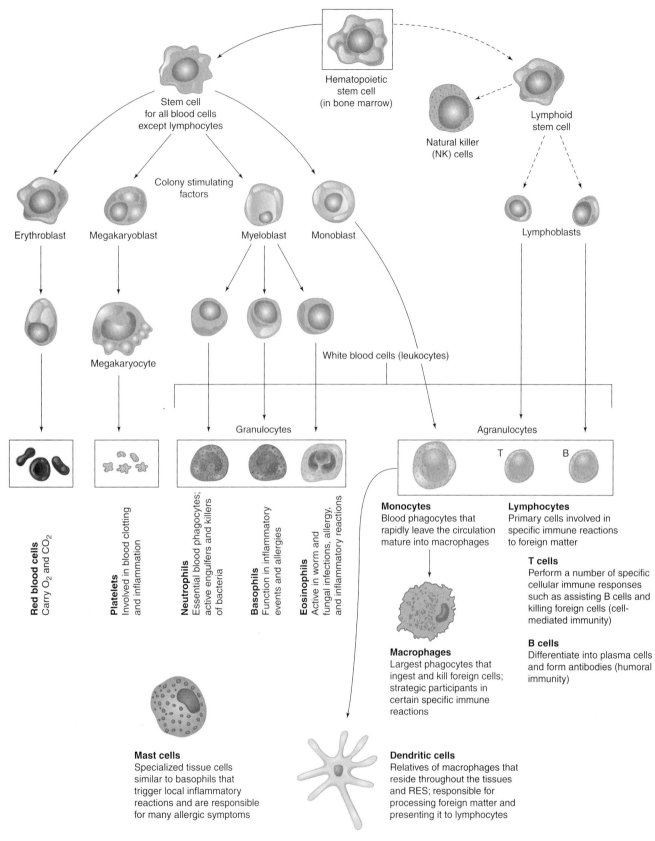

Hematopoietic stem cell (in bone marrow)

Stem cell for all blood cells except lymphocytes

Natural killer (NK) cells

Lymphoid stem cell

Colony stimulating factors

Erythroblast

Megakaryoblast

Myeloblast

Monoblast

Lymphoblasts

Megakaryocyte

White blood cells (leukocytes)

Granulocytes

Agranulocytes

T   B

Red blood cells
Carry O$_2$ and CO$_2$

Platelets
Involved in blood clotting and inflammation

Neutrophils
Essential blood phagocytes; active engulfers and killers of bacteria

Basophils
Function in inflammatory events and allergies

Eosinophils
Active in worm and fungal infections, allergy, and inflammatory reactions

Monocytes
Blood phagocytes that rapidly leave the circulation mature into macrophages

Lymphocytes
Primary cells involved in specific immune reactions to foreign matter

T cells
Perform a number of specific cellular immune responses such as assisting B cells and killing foreign cells (cell-mediated immunity)

B cells
Differentiate into plasma cells and form antibodies (humoral immunity)

Macrophages
Largest phagocytes that ingest and kill foreign cells; strategic participants in certain specific immune reactions

Mast cells
Specialized tissue cells similar to basophils that trigger local inflammatory reactions and are responsible for many allergic symptoms

Dendritic cells
Relatives of macrophages that reside throughout the tissues and RES; responsible for processing foreign matter and presenting it to lymphocytes

The development of blood cells and platelets
Figure 14.9

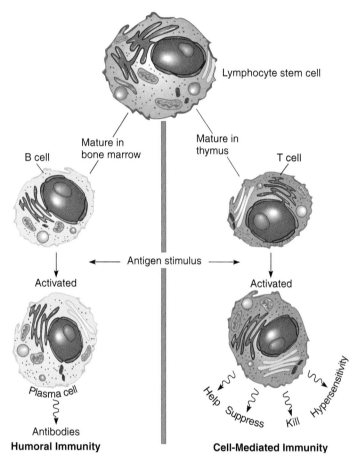

Lymphocyte stem cell

Mature in
bone marrow

Mature in
thymus

B cell

T cell

Antigen stimulus

Activated

Activated

Plasma cell

Help   Suppress   Kill   Hypersensitivity

Antibodies

Humoral Immunity

Cell-Mediated Immunity

**Summary of the general development and
functions of lymphocytes, which are the
cornerstone of specific immune reactions**
Figure 14.10

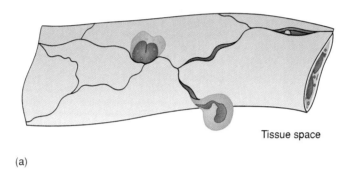

Tissue space

(a)

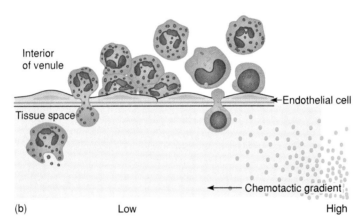

Interior
of venule

Endothelial cell

Tissue space

Chemotactic gradient

(b)   Low   High

Diapedesis and chemotaxis of leukocytes
Figure 14.11

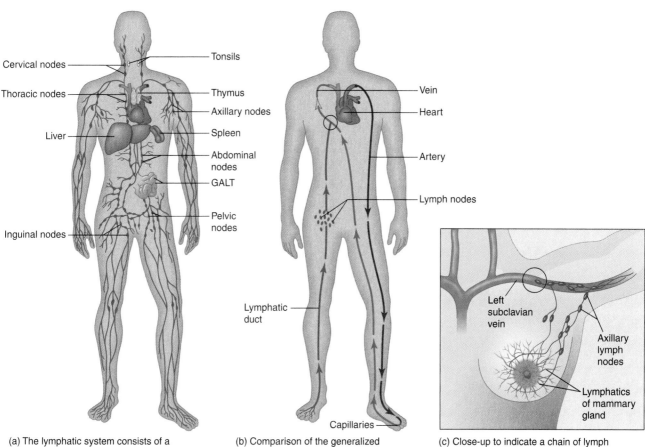

Cervical nodes

Thoracic nodes

Liver

Inguinal nodes

Tonsils

Thymus

Axillary nodes

Spleen

Abdominal nodes

GALT

Pelvic nodes

(a) The lymphatic system consists of a branching network of vessels which extend into most body areas. Note the higher density of lymphatic vessels in the "dead-end" areas of the hands, feet, and breast, which are frequent contact points for infections. Other lymphatic organs include the lymph nodes, spleen, gut-associated lymphoid tissue (GALT), the thymus gland, and the tonsils.

Vein

Heart

Artery

Lymph nodes

Lymphatic duct

Capillaries

(b) Comparison of the generalized circulation of the lymphatic system and the blood. Although the lymphatic vessels parallel the regular circulation, they transport in only one direction unlike the cyclic pattern of blood. Direct connection between the two circulations occurs at points near the heart where large lymph ducts empty their fluid into veins (circled area).

Left subclavian vein

Axillary lymph nodes

Lymphatics of mammary gland

(c) Close-up to indicate a chain of lymph nodes near the axilla and breast and another point of contact between the two circulations (circled area).

General components of the lymphatic system
Figure 14.12

(a) The finest level of lymphatic circulation begins with blind capillaries that pick up fluid, white blood cells, and microbes or other foreign matter from the surrounding tissues and transport this liquid mixture (lymph) away from the extremities via a system of small ducts.

(b) The ducts carry lymph into a circuit of larger ducts that ultimately flow into clusters of specialized filtering organs, the lymph nodes.

(c) The center diagram shows a section through a lymph node to reveal the afferent ducts draining into sinuses that house several types of white blood cells, primarily T lymphocytes, B lymphocytes, macrophages, and dendritic cells. Here, foreign material is filtered out, processed, and becomes the focus of various immune responses.

(d) Lymph continues to trickle from the lymph nodes via efferent ducts into a system of larger drainage vessels, which ultimately connect with large veins near the heart. In this way, cells and products of immunity continually enter the regular circulation.

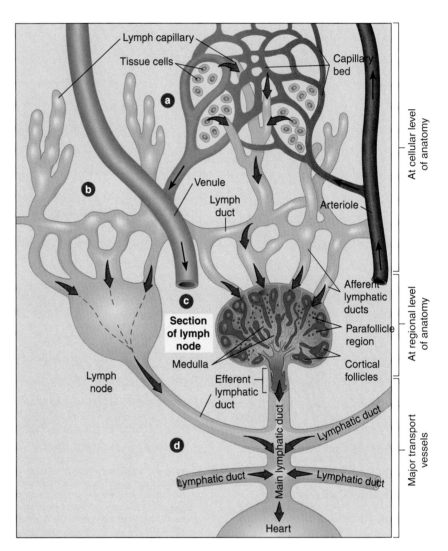

Scheme of circulation in the lymphatic vessels and lymph nodes
Figure 14.13

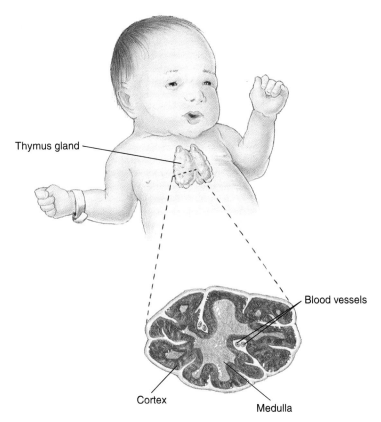

Thymus gland

Blood vessels

Cortex

Medulla

The thymus gland
Figure 14.14

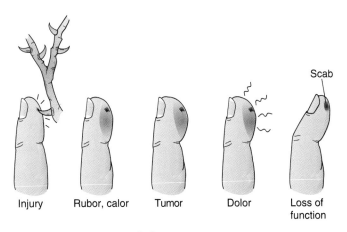

Scab

Injury

Rubor, calor

Tumor

Dolor

Loss of
function

The response to injury
Figure 14.15

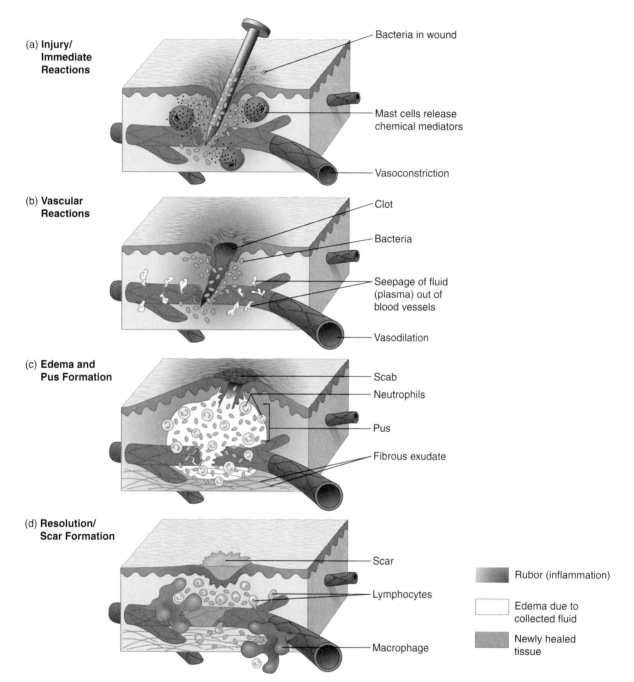

(a) **Injury/ Immediate Reactions**

Bacteria in wound

Mast cells release chemical mediators

Vasoconstriction

(b) **Vascular Reactions**

Clot

Bacteria

Seepage of fluid (plasma) out of blood vessels

Vasodilation

(c) **Edema and Pus Formation**

Scab

Neutrophils

Pus

Fibrous exudate

(d) **Resolution/ Scar Formation**

Scar

Lymphocytes

Macrophage

Rubor (inflammation)

Edema due to collected fluid

Newly healed tissue

The major events in inflammation
Figure 14.16

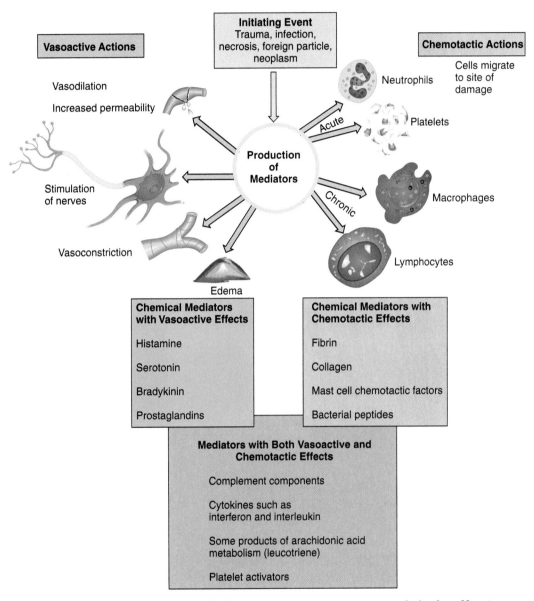

Chemical mediators of the inflammatory response and their effects
Figure 14.17

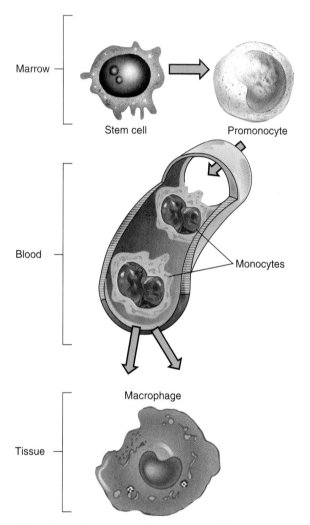

Marrow

Stem cell

Promonocyte

Blood

Monocytes

Macrophage

Tissue

**The developmental stages of monocytes
and macrophages**
Figure 14.18

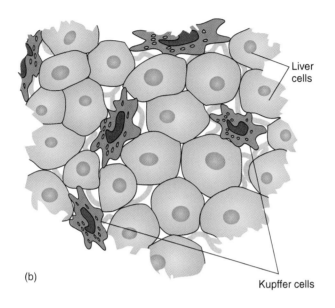

Liver
cells

(b)

Kupffer cells

Langerhans cells

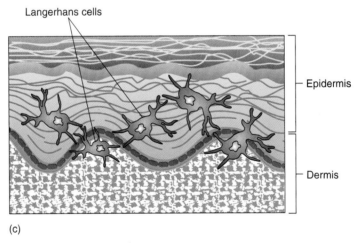

Epidermis

Dermis

(c)

Sites containing macrophages
Figure 14.19

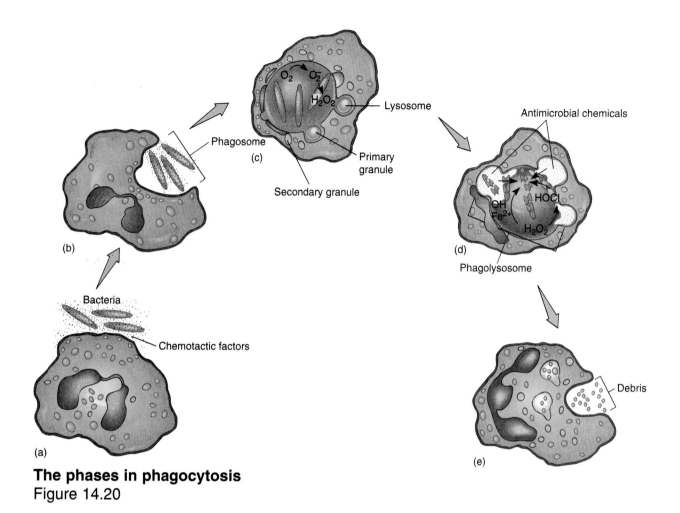

The phases in phagocytosis
Figure 14.20

(a)
Bacteria
Chemotactic factors

(b)

(c)
Phagosome
O_2 O_2^-
H_2O_2
Lysosome
Primary granule
Secondary granule

(d)
Antimicrobial chemicals
OH
Fe^{2+}
HOCl
H_2O_2
Phagolysosome

(e)
Debris

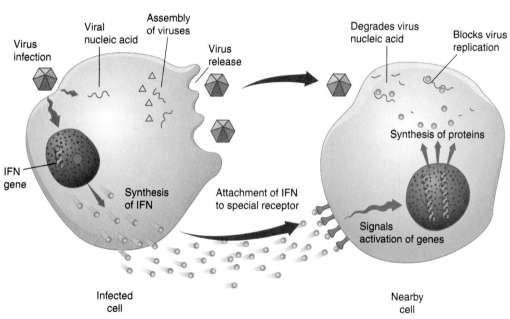

The antiviral activity of interferon
Figure 14.21

Virus infection
Viral nucleic acid
Assembly of viruses
Virus release
IFN gene
Synthesis of IFN
Attachment of IFN to special receptor
Infected cell

Degrades virus nucleic acid
Blocks virus replication
Synthesis of proteins
Signals activation of genes
Nearby cell

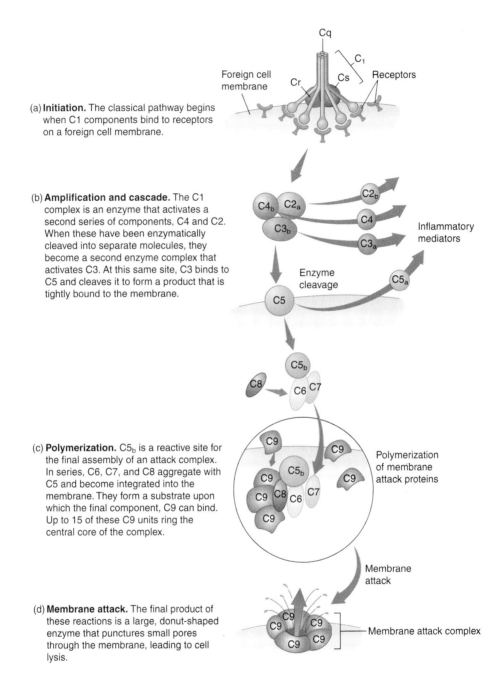

(a) **Initiation.** The classical pathway begins when C1 components bind to receptors on a foreign cell membrane.

(b) **Amplification and cascade.** The C1 complex is an enzyme that activates a second series of components, C4 and C2. When these have been enzymatically cleaved into separate molecules, they become a second enzyme complex that activates C3. At this same site, C3 binds to C5 and cleaves it to form a product that is tightly bound to the membrane.

(c) **Polymerization.** $C5_b$ is a reactive site for the final assembly of an attack complex. In series, C6, C7, and C8 aggregate with C5 and become integrated into the membrane. They form a substrate upon which the final component, C9 can bind. Up to 15 of these C9 units ring the central core of the complex.

(d) **Membrane attack.** The final product of these reactions is a large, donut-shaped enzyme that punctures small pores through the membrane, leading to cell lysis.

Steps in the classical complement pathway at a single site
Figure 14.22

(a) Specificity: Viruses and other infectious agents contain antigen molecules that are specific to a single type of lymphocyte. One result of binding will be the production of virus-specific antibodies.

(b) Memory: First contact with antigen creates a unique programmed memory cell that provides quick recall upon second and other future contacts with that antigen.

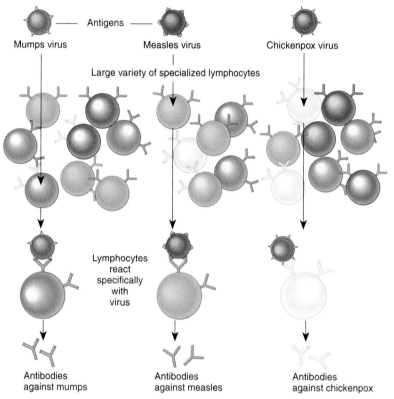

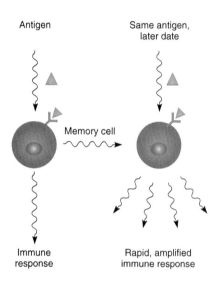

The characteristics of acquired immunity
Figure 14.23

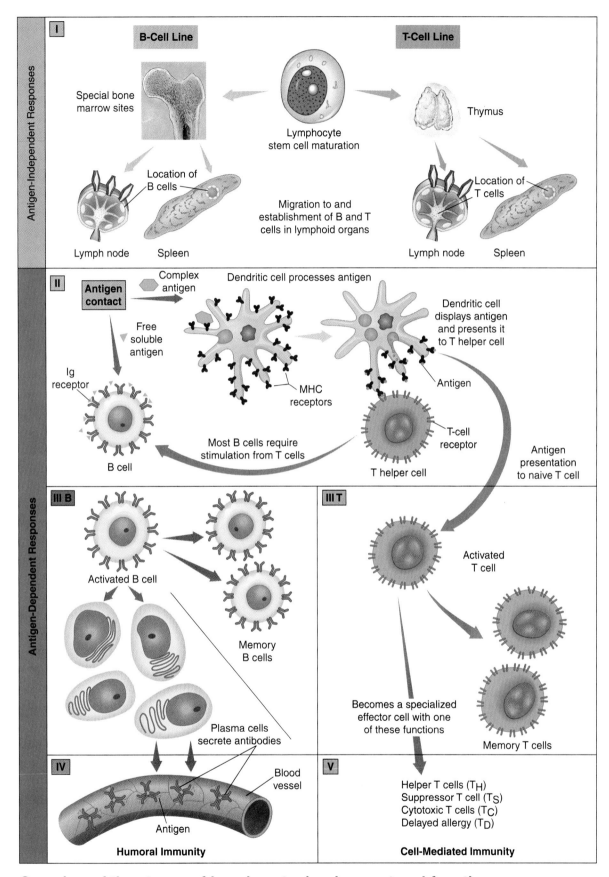

I

Antigen-Independent Responses

B-Cell Line

Special bone marrow sites

Lymphocyte stem cell maturation

T-Cell Line

Thymus

Location of B cells

Lymph node   Spleen

Migration to and establishment of B and T cells in lymphoid organs

Location of T cells

Lymph node   Spleen

II

Antigen-Dependent Responses

Antigen contact

Complex antigen

Dendritic cell processes antigen

Dendritic cell displays antigen and presents it to T helper cell

Free soluble antigen

Ig receptor

MHC receptors

Antigen

Most B cells require stimulation from T cells

B cell

T-cell receptor

T helper cell

Antigen presentation to naive T cell

III B

Activated B cell

Memory B cells

Plasma cells secrete antibodies

III T

Activated T cell

Becomes a specialized effector cell with one of these functions

Memory T cells

IV

Blood vessel

Antigen

Humoral Immunity

V

Helper T cells (T_H)
Suppressor T cell (T_S)
Cytotoxic T cells (T_C)
Delayed allergy (T_D)

Cell-Mediated Immunity

Overview of the stages of lymphocyte development and function
Figure 15.1

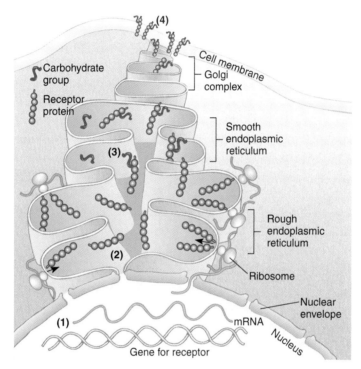

Receptor formation in a developing cell
Figure 15.2

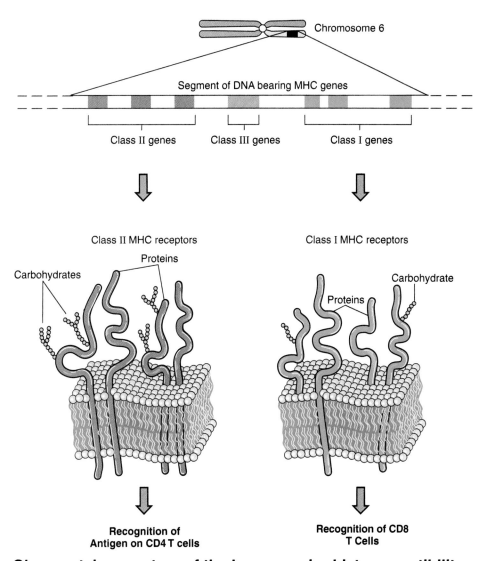

Glycoprotein receptors of the human major histocompatibility (human leukocyte antigen) gene complex (MHC)

Figure 15.3

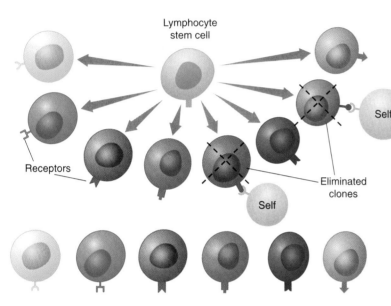

Lymphocyte
stem cell

Receptors

Self

Eliminated
clones

Self

Repertoire of lymphocyte clones, each with unique receptor display

(a) **Antigen-Independent Period**
1. During development of early lymphocytes from stem cells, a given stem cell undergoes rapid cell division to form numerous progeny.

During this period of cell differentiation, random rearrangements of the genes that code for cell surface protein receptors occur. The result is a large array of genetically distinct cells, called clones, each clone bearing a different receptor that is specific to react with only a single type of foreign molecule or antigen.

2. At this same time, any lymphocyte clones that develop a specificity for self molecules and could be harmful are eliminated or deleted from the pool of diversity. This is called immune tolerance.

3. The specificity for a single antigen molecule is programmed into the lymphocyte and is set for the life of a given clone. The end result is an enormous pool of immature or naive lymphocytes that are ready to further differentiate under the influence of certain organs and immune stimuli.

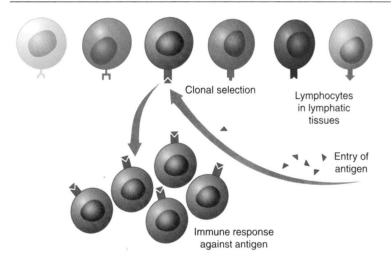

Clonal selection

Lymphocytes
in lymphatic
tissues

Entry of
antigen

Immune response
against antigen

(b) **Antigen-Dependent Period**
4. Lymphocytes come to populate the lymphatic organs, where they will finally encounter antigens. These antigens will become the stimulus for the lymphocytes' final activation and immune function. Entry of a specific antigen selects only the lymphocyte clone or clones that carries matching surface receptors. This will trigger an immune response, which varies according to the type of lymphocyte involved.

Overview of the clonal selection theory of lymphocyte development and diversity
Figure 15.4

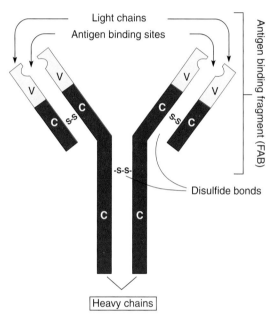

Simplified structure of an immunoglobulin molecule
Figure 15.5

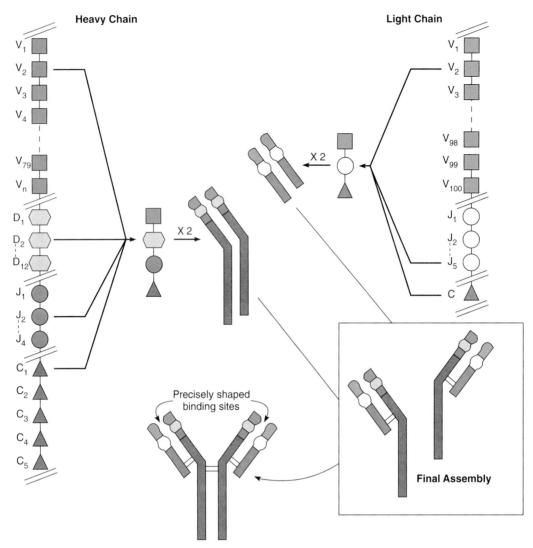

A simplified look at immunoglobulin genetics
Figure 15.6

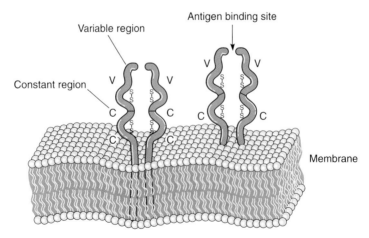

Proposed structure of the T-cell receptor for antigen
Figure 15.7

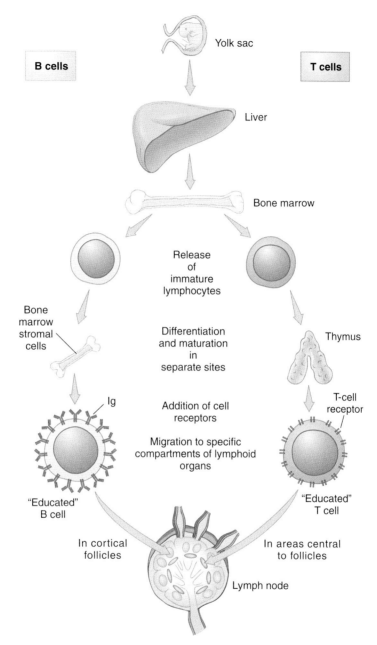

Yolk sac

B cells

T cells

Liver

Bone marrow

Release
of
immature
lymphocytes

Bone
marrow
stromal
cells

Differentiation
and maturation
in
separate sites

Thymus

Ig

Addition of cell
receptors

T-cell
receptor

Migration to specific
compartments of lymphoid
organs

"Educated"
B cell

"Educated"
T cell

In cortical
follicles

In areas central
to follicles

Lymph node

**Major stages in the development of B and
T cells**
Figure 15.8

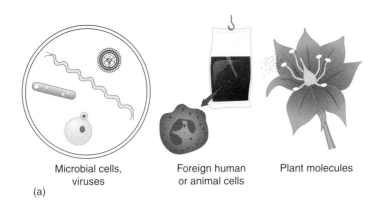

Microbial cells, viruses

Foreign human or animal cells

Plant molecules

(a)

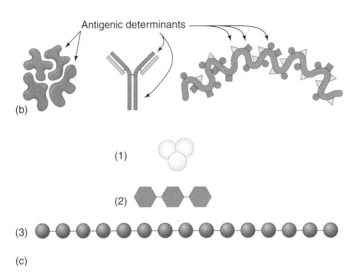

Antigenic determinants

(b)

(1)

(2)

(3)

(c)

Characteristics of antigens
Figure 15.9

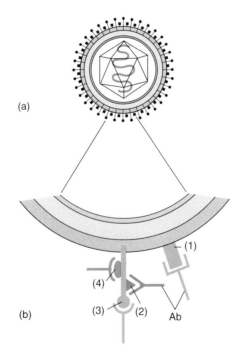

(a)

(1)

(4)

(3) (2)

Ab

(b)

Mosaic antigens
Figure 15.10

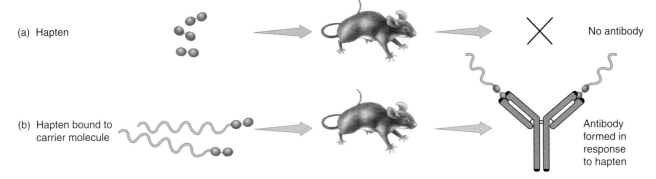

(a) Hapten → → ✗ No antibody

(b) Hapten bound to carrier molecule → → Antibody formed in response to hapten

The hapten-carrier phenomenon
Figure 15.11

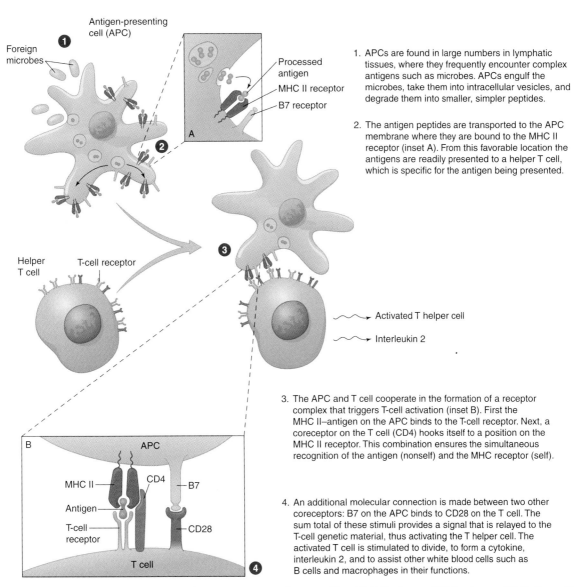

Foreign microbes

Antigen-presenting cell (APC) ❶

Processed antigen
MHC II receptor
B7 receptor

A

❷

Helper T cell

T-cell receptor

❸

Activated T helper cell

Interleukin 2

B
APC
MHC II
CD4
B7
Antigen
T-cell receptor
CD28
T cell
❹

1. APCs are found in large numbers in lymphatic tissues, where they frequently encounter complex antigens such as microbes. APCs engulf the microbes, take them into intracellular vesicles, and degrade them into smaller, simpler peptides.

2. The antigen peptides are transported to the APC membrane where they are bound to the MHC II receptor (inset A). From this favorable location the antigens are readily presented to a helper T cell, which is specific for the antigen being presented.

3. The APC and T cell cooperate in the formation of a receptor complex that triggers T-cell activation (inset B). First the MHC II–antigen on the APC binds to the T-cell receptor. Next, a coreceptor on the T cell (CD4) hooks itself to a position on the MHC II receptor. This combination ensures the simultaneous recognition of the antigen (nonself) and the MHC receptor (self).

4. An additional molecular connection is made between two other coreceptors: B7 on the APC binds to CD28 on the T cell. The sum total of these stimuli provides a signal that is relayed to the T-cell genetic material, thus activating the T helper cell. The activated T cell is stimulated to divide, to form a cytokine, interleukin 2, and to assist other white blood cells such as B cells and macrophages in their functions.

Interactions between antigen-presenting cells (APCs) and T helper (CD4) cells required for T-cell activation
Figure 15.12

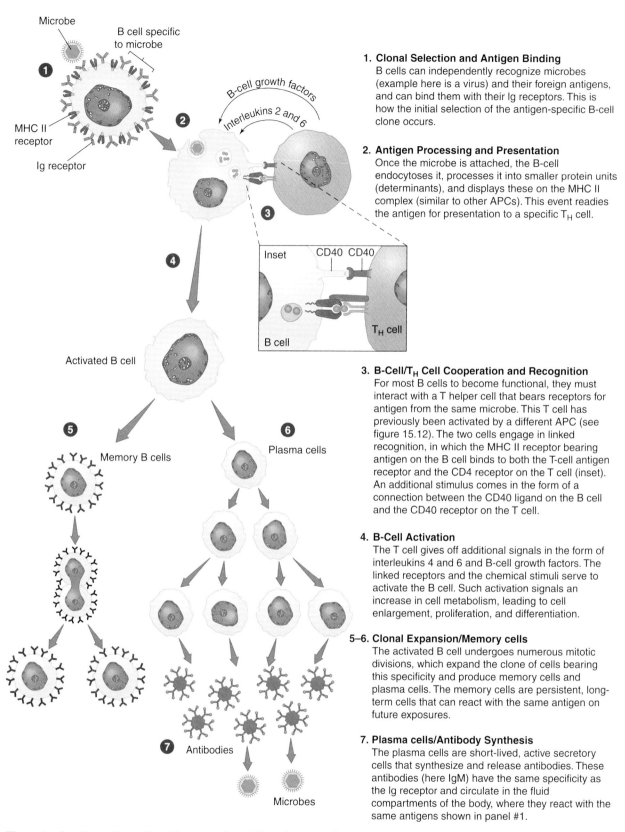

1. **Clonal Selection and Antigen Binding**
   B cells can independently recognize microbes (example here is a virus) and their foreign antigens, and can bind them with their Ig receptors. This is how the initial selection of the antigen-specific B-cell clone occurs.

2. **Antigen Processing and Presentation**
   Once the microbe is attached, the B-cell endocytoses it, processes it into smaller protein units (determinants), and displays these on the MHC II complex (similar to other APCs). This event readies the antigen for presentation to a specific T_H cell.

3. **B-Cell/T_H Cell Cooperation and Recognition**
   For most B cells to become functional, they must interact with a T helper cell that bears receptors for antigen from the same microbe. This T cell has previously been activated by a different APC (see figure 15.12). The two cells engage in linked recognition, in which the MHC II receptor bearing antigen on the B cell binds to both the T-cell antigen receptor and the CD4 receptor on the T cell (inset). An additional stimulus comes in the form of a connection between the CD40 ligand on the B cell and the CD40 receptor on the T cell.

4. **B-Cell Activation**
   The T cell gives off additional signals in the form of interleukins 4 and 6 and B-cell growth factors. The linked receptors and the chemical stimuli serve to activate the B cell. Such activation signals an increase in cell metabolism, leading to cell enlargement, proliferation, and differentiation.

5–6. **Clonal Expansion/Memory cells**
   The activated B cell undergoes numerous mitotic divisions, which expand the clone of cells bearing this specificity and produce memory cells and plasma cells. The memory cells are persistent, long-term cells that can react with the same antigen on future exposures.

7. **Plasma cells/Antibody Synthesis**
   The plasma cells are short-lived, active secretory cells that synthesize and release antibodies. These antibodies (here IgM) have the same specificity as the Ig receptor and circulate in the fluid compartments of the body, where they react with the same antigens shown in panel #1.

Events in B-cell activation and antibody synthesis
Figure 15.13

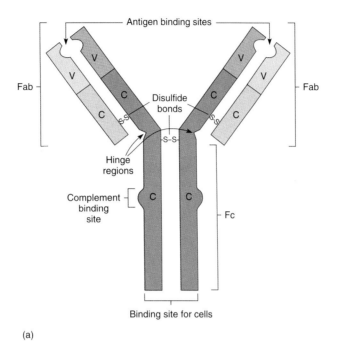

(a)

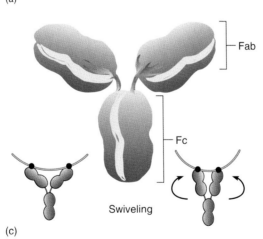

(c)

Working models of antibody structure
Figure 15.14

219

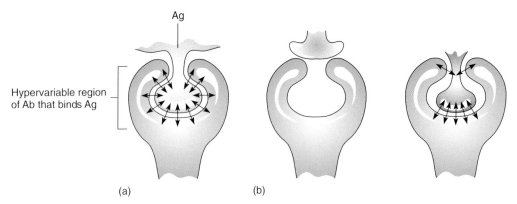

Antigen-antibody binding
Figure 15.15

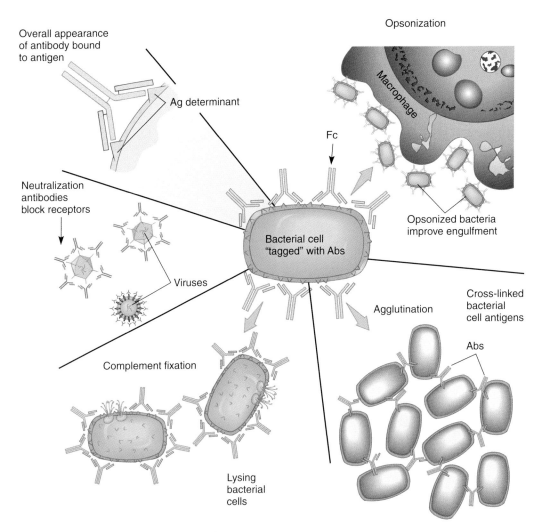

Summary of antibody functions
Figure 15.16

TABLE 15.2

Characteristics of the Immunoglobulin (Ig) Classes

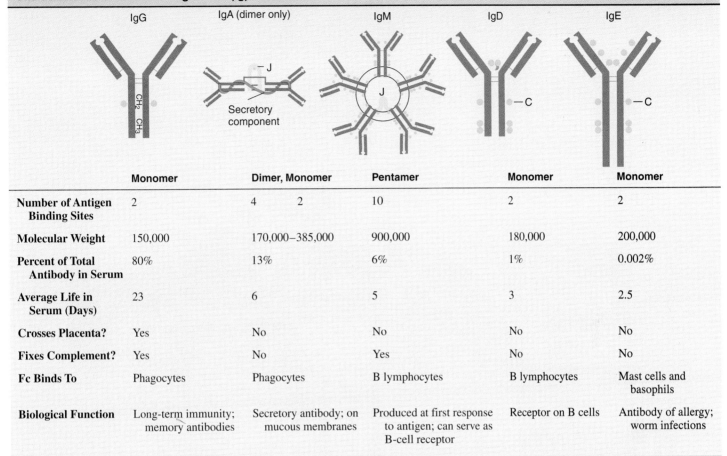

	IgG	IgA (dimer only)	IgM	IgD	IgE
	Monomer	Dimer, Monomer	Pentamer	Monomer	Monomer
Number of Antigen Binding Sites	2	4 2	10	2	2
Molecular Weight	150,000	170,000–385,000	900,000	180,000	200,000
Percent of Total Antibody in Serum	80%	13%	6%	1%	0.002%
Average Life in Serum (Days)	23	6	5	3	2.5
Crosses Placenta?	Yes	No	No	No	No
Fixes Complement?	Yes	No	Yes	No	No
Fc Binds To	Phagocytes	Phagocytes	B lymphocytes	B lymphocytes	Mast cells and basophils
Biological Function	Long-term immunity; memory antibodies	Secretory antibody; on mucous membranes	Produced at first response to antigen; can serve as B-cell receptor	Receptor on B cells	Antibody of allergy; worm infections

C = carbohydrate.
J = J chain.

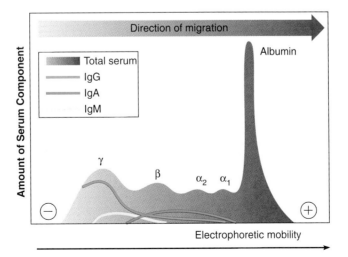

Pattern of human serum after electrophoresis

Figure 15.17

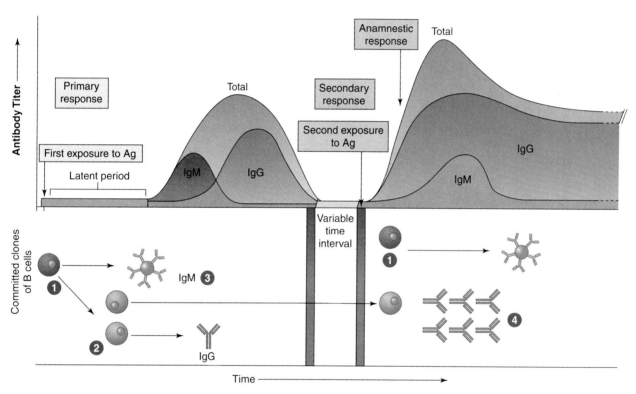

Primary and secondary responses to antigens
Figure 15.18

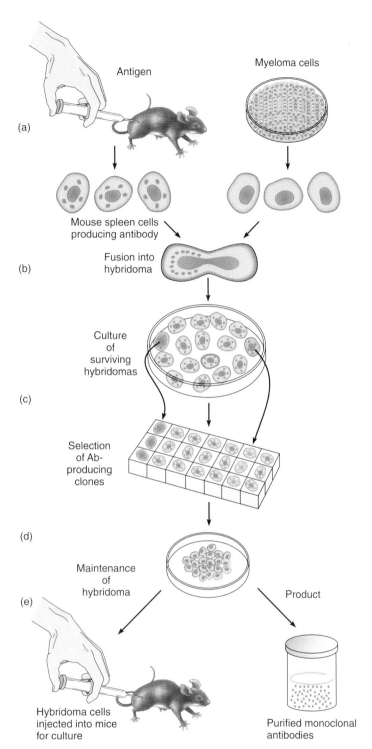

(a)

Antigen

Myeloma cells

Mouse spleen cells producing antibody

(b) Fusion into hybridoma

(c) Culture of surviving hybridomas

Selection of Ab-producing clones

(d)

(e) Maintenance of hybridoma

Product

Hybridoma cells injected into mice for culture

Purified monoclonal antibodies

Summary of the technique for producing monoclonal antibodies by hybridizing myeloma tumor cells with normal plasma cells

Figure 15.19

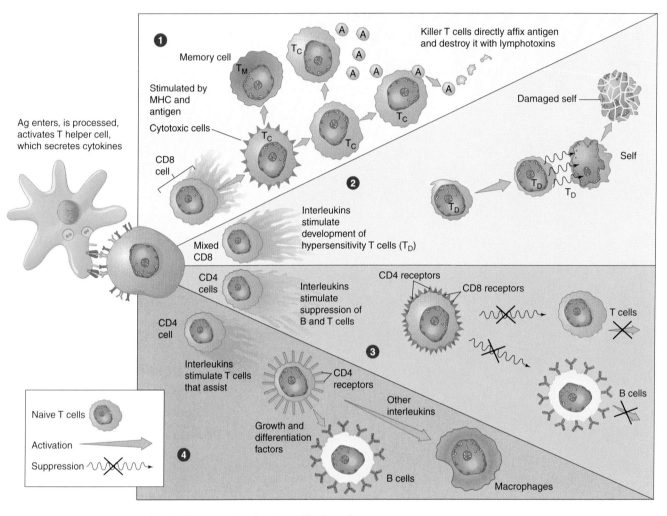

Ag enters, is processed, activates T helper cell, which secretes cytokines

1
Memory cell
T_M
Stimulated by MHC and antigen
Cytotoxic cells
CD8 cell
T_C
T_C
T_C
A A A A A A A A A A
Killer T cells directly affix antigen and destroy it with lymphotoxins
Damaged self
Self

2
Interleukins stimulate development of hypersensitivity T cells (T_D)
T_D
T_D
T_D

Mixed CD8

CD4 cells
CD4 cell
Interleukins stimulate suppression of B and T cells
CD4 receptors
CD8 receptors
3
T cells
B cells

Interleukins stimulate T cells that assist
CD4 receptors
Other interleukins
Growth and differentiation factors
4
B cells
Macrophages

Naive T cells
Activation
Suppression

1. T_C cells destroy certain microbes and foreign cells.
2. T_D cells react with allergens and cause a type of hypersensitivity that damages self.
3. T_S cells limit the actions of B and T cells.
4. T_H cells assist in the actions of B and T cells.

Overall scheme of T-cell activation and differentiation into different types of T cells
Figure 15.20

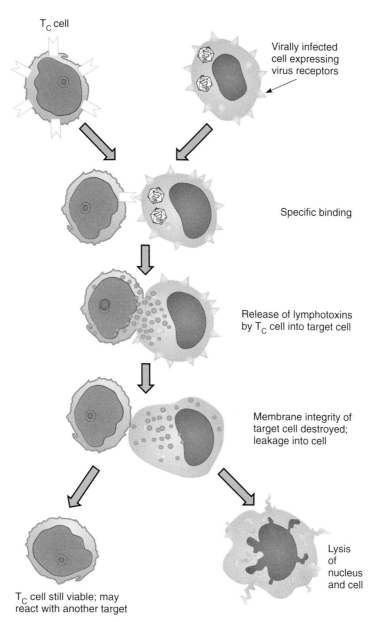

T$_C$ cell

Virally infected cell expressing virus receptors

Specific binding

Release of lymphotoxins by T$_C$ cell into target cell

Membrane integrity of target cell destroyed; leakage into cell

Lysis of nucleus and cell

T$_C$ cell still viable; may react with another target

Stages of cell-mediated cytotoxicity and the action of lymphotoxins on virus-infected target cells
Figure 15.21

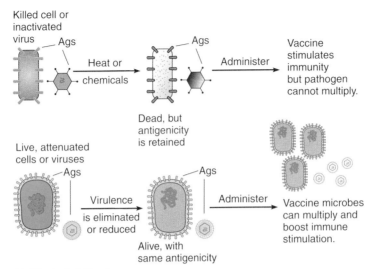

Killed cell or inactivated virus

Ags

Heat or chemicals

Dead, but antigenicity is retained

Ags

Administer

Vaccine stimulates immunity but pathogen cannot multiply.

Live, attenuated cells or viruses

Ags

Virulence is eliminated or reduced

Alive, with same antigenicity

Ags

Administer

Vaccine microbes can multiply and boost immune stimulation.

(a) **Whole Cell Vaccines**

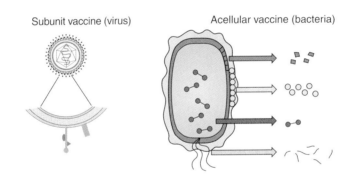

Subunit vaccine (virus)

Acellular vaccine (bacteria)

(b) **Vaccines from Microbe Parts**

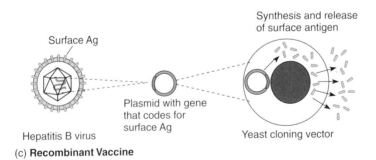

Surface Ag

Hepatitis B virus

Plasmid with gene that codes for surface Ag

Synthesis and release of surface antigen

Yeast cloning vector

(c) **Recombinant Vaccine**

Strategies in vaccine design
Figure 16.1

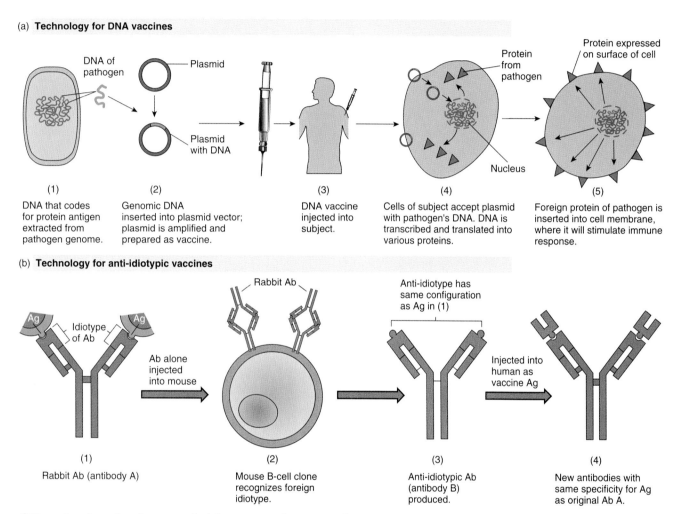

(a) **Technology for DNA vaccines**

DNA of pathogen

Plasmid

Plasmid with DNA

Protein from pathogen

Protein expressed on surface of cell

Nucleus

(1)
DNA that codes for protein antigen extracted from pathogen genome.

(2)
Genomic DNA inserted into plasmid vector; plasmid is amplified and prepared as vaccine.

(3)
DNA vaccine injected into subject.

(4)
Cells of subject accept plasmid with pathogen's DNA. DNA is transcribed and translated into various proteins.

(5)
Foreign protein of pathogen is inserted into cell membrane, where it will stimulate immune response.

(b) **Technology for anti-idiotypic vaccines**

Rabbit Ab

Ag

Idiotype of Ab

Ag

Anti-idiotype has same configuration as Ag in (1)

Ab alone injected into mouse

Injected into human as vaccine Ag

(1)
Rabbit Ab (antibody A)

(2)
Mouse B-cell clone recognizes foreign idiotype.

(3)
Anti-idiotypic Ab (antibody B) produced.

(4)
New antibodies with same specificity for Ag as original Ab A.

Other technologies useful in preparing vaccines
Figure 16.2

TABLE 16.3

Recommended Regimen and Indications for Routine Vaccinations

Vaccine / Age ▶	Birth	1 mo	2 mos	4 mos	6 mos	12 mos	15 mos	18 mos	24 mos	4–6 Yrs	11–12 Yrs	13–18 Yrs	Comments
		range of recommended ages				*catch-up vaccination*				*preadolescent assessment*			
Hepatitis B	HB #1	only if mother HBsAg (–)	HepB #2			HepB #3				HepB series			Option depends upon condition of infant; 3 doses given
Diphtheria, Tetanus, Pertussis[1]			DTaP	DTaP	DTaP		DTaP			DTaP	Td		Td is tetanus/diphtheria; T is tetanus alone; either one should be given as booster every 10 years.
Haemophilus influenzae Type b			Hib	Hib	Hib	Hib							Schedule depends upon source of vaccine; given with DTaP asTriHIBit or with IPV and HB as Pediatrix
Inactivated Polio[2]			IPV	IPV	IPV				IPV				Similar schedule to DTaP; injected vaccine
Measles, Mumps, Rubella[3]						MMR #1			MMR #2	MMR #2			First dose varies with disease incidence; booster given at either 4–6 or 11–12 years
Varicella						Varicella			Varicella				Used to prevent or limit disease severity of chickenpox in susceptible persons
Pneumococcal[4]			PCV	PCV	PCV	PCV			PCV	PPV			Used to protect children against otitis media
Vaccines below this line are for selected populations													
Hepatitis A									Hepatitis A series				
Influenza					Influenza (yearly)								

[1]DTaP. The diphtheria–tetanus–acellular pertussis vaccine has replaced the DTP.

[4]PCV is pneumococcal conjugate vaccine.
PPV is pneumococcal polysaccharide vaccine.
Infants require the more immunogenic conjugate vaccine.

[2]polio vaccine—is the recommended vaccine for all schedules

[3]Measles vaccine (Attenuvax) can be given alone to children during epidemics or to adults immunized before 1970.

Used in Cases of Specific Risk Due to Occupational or Other Exposure

Vaccine	Group Targeted
Hepatitis B (Recombivax)	Health care personnel; people exposed through life-style
Hepatitis A (Havrix)	Children 2–14 years who live in areas of high prevalence
Pneumococcus (Pneumovax)	Elderly patients, children with sickle-cell anemia
Influenza, polio, tuberculosis (BCG)	Hospital, laboratory, health care workers
Rabies, plague, tularemia	People whose jobs involve contact with animals (veterinarians, forest rangers); known or suspected exposure to rabid animal; living in areas of high incidence
Cholera, hepatitis B, hepatitis A, measles, yellow fever, meningococcal meningitis, polio, rabies, typhoid, plague, botulism	Travelers to endemic regions, including military recruits (varies with geographic destination), laboratory workers

This schedule indicates the recommended ages for routine administration of currently licensed childhood vaccines, as of December 1, 2002, for children through age 18 years. Any dose not given at the recommended age should be given at any subsequent visit when indicated and feasible. ▨ Indicates age groups that warrant special effort to administer those vaccines not previously given. Additional vaccines may be licensed and recommended during the year. Licensed combination vaccines may be used whenever any components of the combination are indicated and the vaccine's other components are not contraindicated. Providers should consult the manufacturers' package inserts for detailed recommendations.

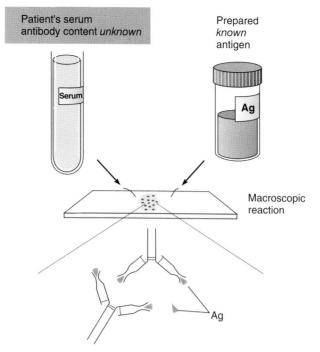

Patient's serum
antibody content *unknown*

Prepared
known
antigen

Serum

Ag

Macroscopic
reaction

Ag

(a) In serological diagnosis of disease, a blood sample is scanned for
the presence of antibody using an antigen of known specificity. A
positive reaction is usually evident as some visible sign, such as
color change or clumping, that indicates a specific interaction
between antibody and antigen. (The reaction at the molecular level
is rarely observed.)

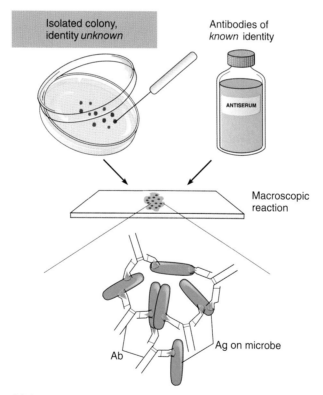

Isolated colony,
identity *unknown*

Antibodies of
known identity

ANTISERUM

Macroscopic
reaction

Ab

Ag on microbe

(b) An unknown microbe is mixed with serum containing antibodies
of known specificity, a procedure known as serotyping.
Microscopically or macroscopically observable reactions indicate
a correct match between antibody and antigen and permit
identification of the microbe.

Basic principles of serological testing using antibodies and antigens
Figure 16.3

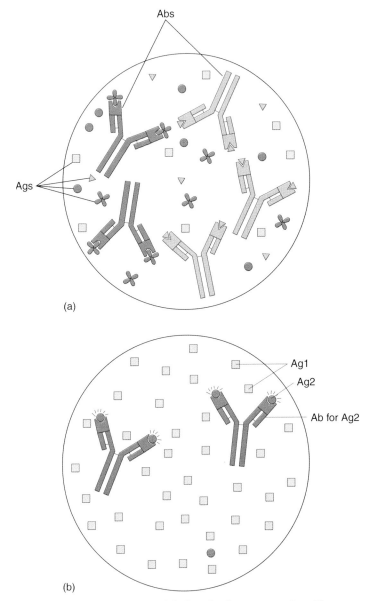

Specificity and sensitivity in immune testing
Figure 16.4

Agglutination

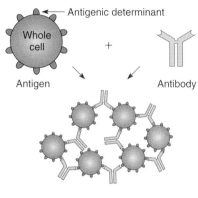

Antigenic determinant

Whole cell

+

Antigen   Antibody

Microscopic appearance of clumps

Precipitation

Cell-free molecule in solution

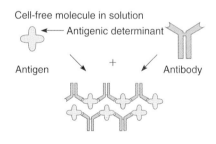

Antigenic determinant

Antigen   +   Antibody

Microscopic appearance of precipitate

(a)

The Tube Agglutination Test

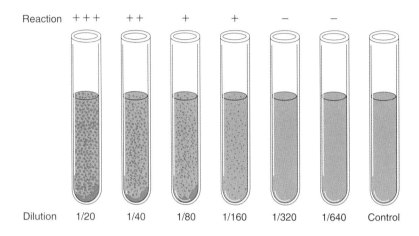

Reaction	+++	++	+	+	−	−	
Dilution	1/20	1/40	1/80	1/160	1/320	1/640	Control

A sample of patient's serum is serially diluted with saline. The dilution is made in a way that halves the number of antibodies in each subsequent tube. An equal amount of the antigen (here, blue bacterial cells) is added to each tube. The control tube has antigen, but no serum. After incubation and centrifugation, each tube is examined for agglutination clumps as compared with the control, which will be cloudy and clump-free. The titer is defined as the dilution of the last tube in the series that shows agglutination.

(b)

Cellular/molecular view of agglutination and precipitation reactions that produce visible antigen-antibody complexes
Figure 16.5

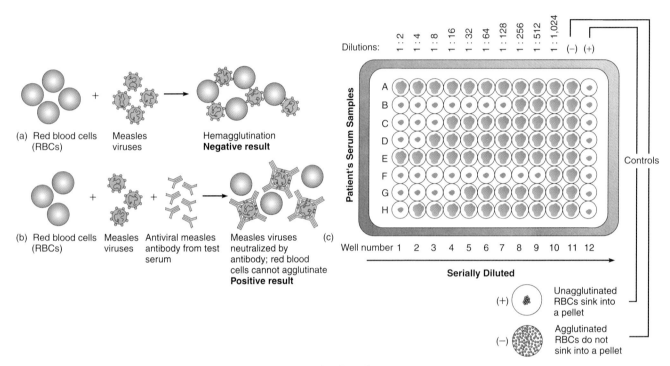

Theory and interpretation of viral hemagglutination
Figure 16.6

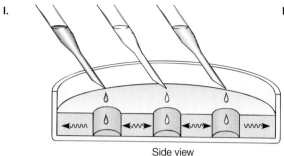

I. In one method of setting up a double-diffusion test, wells are punctured in soft agar, and antibodies (Ab) and antigens (Ag) are added in a pattern. As the contents of the wells diffuse toward each other, a number of reactions can result, depending on whether antibodies meet and precipitate antigens.

Side view

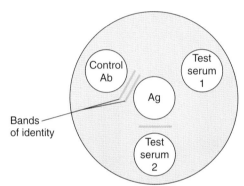

II. Example of test pattern and results. Antigen (Ag) is placed in the center well and antibody (Ab) samples are placed in outer wells. The control contains known Abs to the test Ag. Note bands that form where Ab/Ag meet. The other wells (1, 2) contain unknown test sera. One is positive and the other is negative. Double bands indicate more than one antigen and antibody that can react.

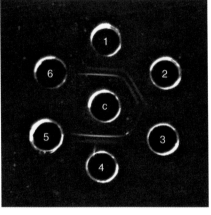

III. Actual test results for detecting infection with the fungal pathogen *Histoplasma*. Numbers 1 and 4 are controls and 2, 3, 5, 6 are patient test sera. Determine which patients have the infection and which do not.

(b)

Precipitation reactions

Figure 16.7

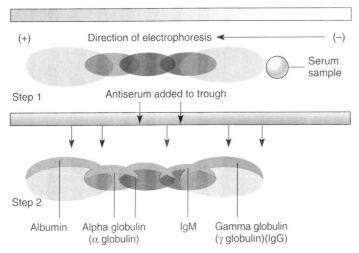

Immunoelectrophoresis of normal human serum
Figure 16.8

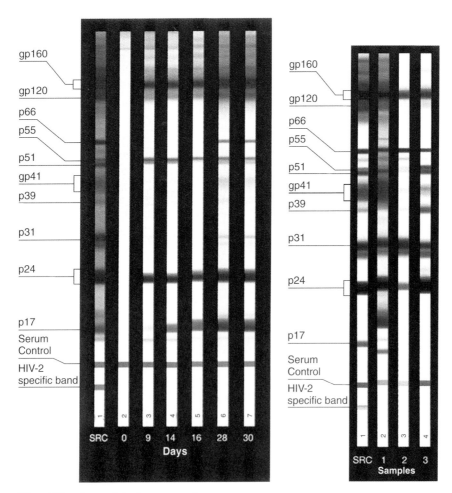

The Western blot procedure
Figure 16.9

Genelabs Diagnostics Pte Ltd

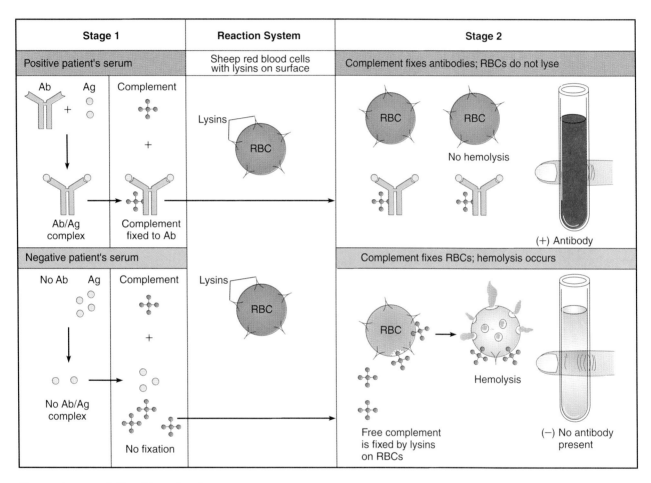

Complement fixation test

Figure 16.10

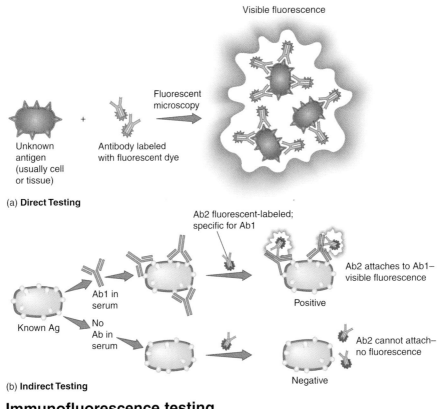

Visible fluorescence

Unknown
antigen
(usually cell
or tissue)

+

Antibody labeled
with fluorescent dye

Fluorescent
microscopy

(a) **Direct Testing**

Ab2 fluorescent-labeled;
specific for Ab1

Ab1 in
serum

Known Ag

No
Ab in
serum

Ab2 attaches to Ab1–
visible fluorescence

Positive

Ab2 cannot attach–
no fluorescence

Negative

(b) **Indirect Testing**

Immunofluorescence testing
Figure 16.11

(a) **Indirect ELISA,** comparing a positive vs negative reaction. This is the basis for HIV screening tests.

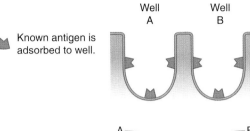

Known antigen is adsorbed to well.

Well A   Well B

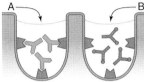

Serum samples with unknown antibodies.

A

B

Well is washed to remove unbound (nonreactive) antibodies.

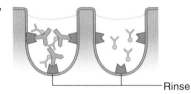

Indicator antibody linked to enzyme attaches to any bound antibody.

Rinse

Wells are rinsed to remove unbound indicator antibody; A colorless substrate for enzyme is added.

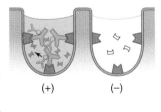

Enzymes linked to indicator Ab hydrolyze the substrate, which releases a dye. Wells that develop color are positive for the antibody: colorless wells are negative.

(+)   (−)

Methods of ELISA testing
Figure 16.12

(c) **Capture or Antibody Sandwich ELISA method.**
Note that an antigen is trapped between two antibodies. This test is used to detect hantavirus and measles virus.

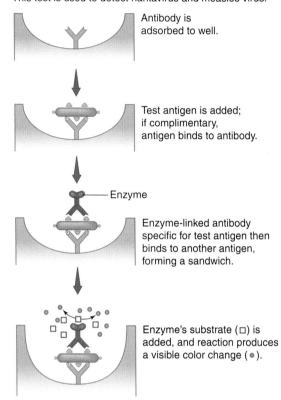

Antibody is adsorbed to well.

Test antigen is added; if complimentary, antigen binds to antibody.

Enzyme

Enzyme-linked antibody specific for test antigen then binds to another antigen, forming a sandwich.

Enzyme's substrate (□) is added, and reaction produces a visible color change (•).

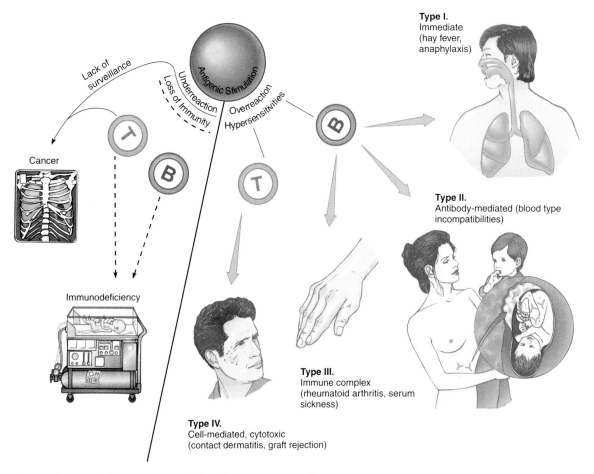

Type I.
Immediate (hay fever, anaphylaxis)

Type II.
Antibody-mediated (blood type incompatibilities)

Type III.
Immune complex (rheumatoid arthritis, serum sickness)

Type IV.
Cell-mediated, cytotoxic (contact dermatitis, graft rejection)

Lack of surveillance

Loss of Immunity

Underreaction

Antigenic Stimulation

Overreaction

Hypersensitivities

Cancer

Immunodeficiency

Overview of diseases of the immune system
Figure 17.1

National Allergy Bureau
Pollen and Mold Report

Location: Sacramento, CA   Date: June 04, 2003
Counting Station: Allergy Medical Group of the North Area

Trees	Moderate severity	Total count: 41 / m^3
Weeds	High severity	Total count: 64 / m^3
Grass	High severity	Total count: 60 / m^3
Mold	Low severity	Total count: 4,219 / m^3

(a)

Monitoring airborne allergens
Figure 17.2

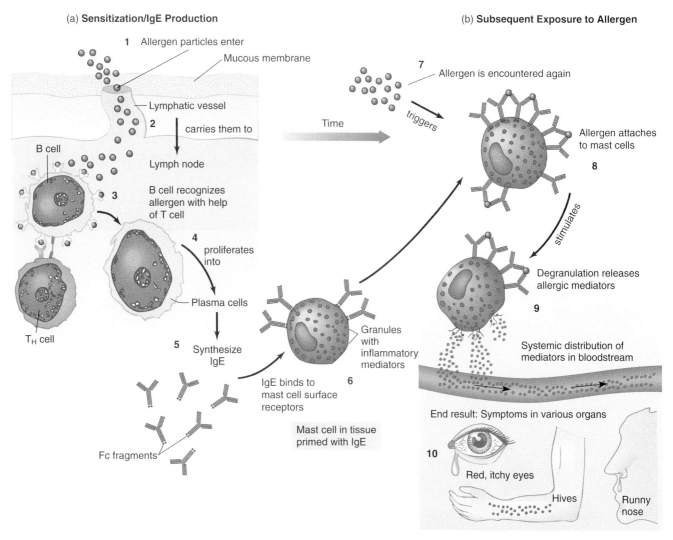

(a) **Sensitization/IgE Production**

1 Allergen particles enter

Mucous membrane

Lymphatic vessel

2 carries them to

Lymph node

B cell

3 B cell recognizes allergen with help of T cell

4 proliferates into

Plasma cells

T_H cell

5 Synthesize IgE

Fc fragments

6 IgE binds to mast cell surface receptors

Granules with inflammatory mediators

Mast cell in tissue primed with IgE

(b) **Subsequent Exposure to Allergen**

Time

7 Allergen is encountered again

triggers

Allergen attaches to mast cells

8

stimulates

Degranulation releases allergic mediators

9

Systemic distribution of mediators in bloodstream

End result: Symptoms in various organs

10

Red, itchy eyes

Hives

Runny nose

A schematic view of cellular reactions during the type I allergic response
Figure 17.3

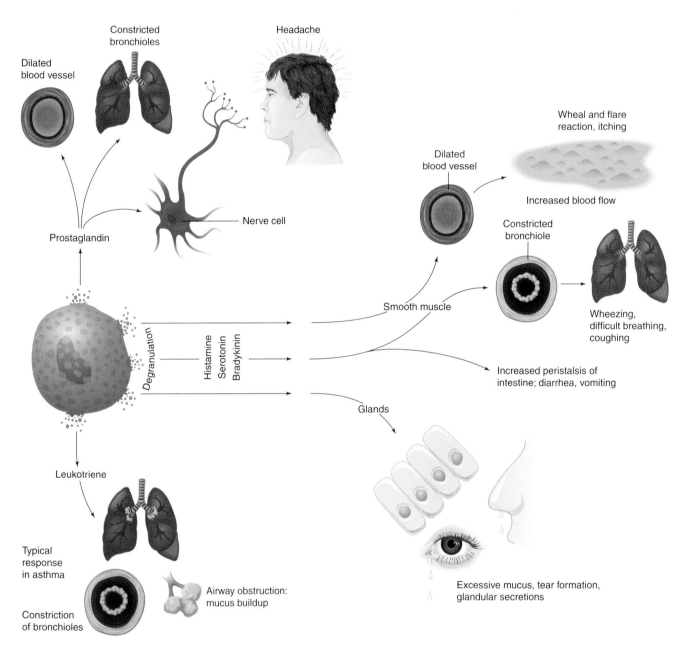

The spectrum of reactions to inflammatory cytokines released by mast cells and the common symptoms they elicit in target tissues and organs

Figure 17.4

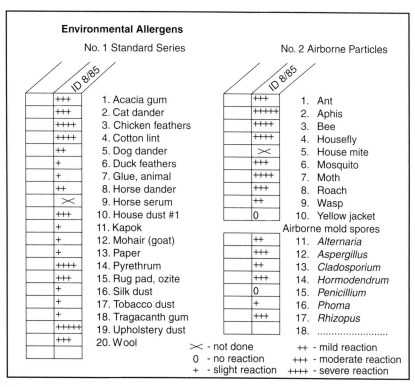

Environmental Allergens

No. 1 Standard Series

ID 8/85	
+++	1. Acacia gum
+++	2. Cat dander
++++	3. Chicken feathers
++++	4. Cotton lint
++	5. Dog dander
+	6. Duck feathers
+	7. Glue, animal
++	8. Horse dander
✕	9. Horse serum
+++	10. House dust #1
+	11. Kapok
+	12. Mohair (goat)
+	13. Paper
++++	14. Pyrethrum
+++	15. Rug pad, ozite
+	16. Silk dust
+	17. Tobacco dust
+	18. Tragacanth gum
+++++	19. Upholstery dust
+++	20. Wool

No. 2 Airborne Particles

ID 8/85	
+++	1. Ant
+++++	2. Aphis
++++	3. Bee
++++	4. Housefly
✕	5. House mite
+++	6. Mosquito
++++	7. Moth
+++	8. Roach
++	9. Wasp
0	10. Yellow jacket

Airborne mold spores

++	11. *Alternaria*
+++	12. *Aspergillus*
++	13. *Cladosporium*
+++	14. *Hormodendrum*
0	15. *Penicillium*
+	16. *Phoma*
+++	17. *Rhizopus*
	18. ....

✕ - not done
0 - no reaction
+ - slight reaction
++ - mild reaction
+++ - moderate reaction
++++ - severe reaction

(b)

A method for conducting an allergy skin test
Figure 17.6

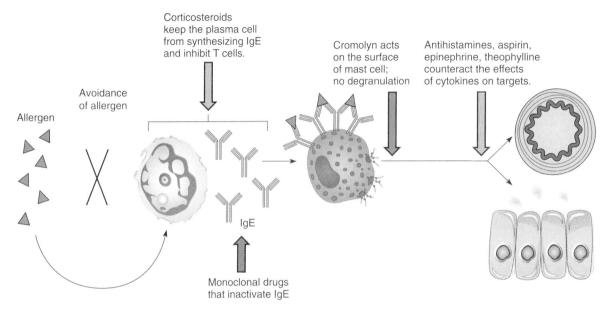

Corticosteroids keep the plasma cell from synthesizing IgE and inhibit T cells.

Cromolyn acts on the surface of mast cell; no degranulation

Antihistamines, aspirin, epinephrine, theophylline counteract the effects of cytokines on targets.

Avoidance of allergen

Allergen

IgE

Monoclonal drugs that inactivate IgE

Strategies for circumventing allergic attacks
Figure 17.7

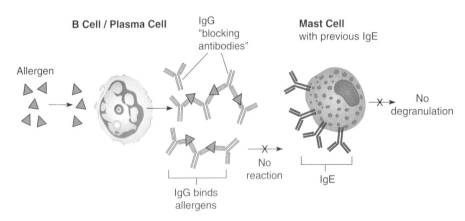

B Cell / Plasma Cell

IgG "blocking antibodies"

Mast Cell with previous IgE

Allergen

No degranulation

No reaction

IgE

IgG binds allergens

The blocking antibody theory for allergic desensitization
Figure 17.8

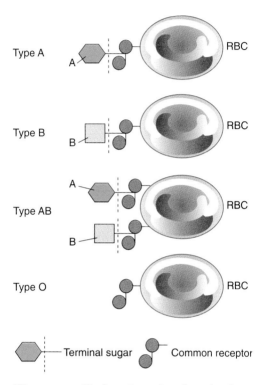

Type A

A

RBC

Type B

B

RBC

Type AB

A

B

RBC

Type O

RBC

Terminal sugar   Common receptor

The genetic/molecular basis for the A and B antigens (receptors) on red blood cells
Figure 17.9

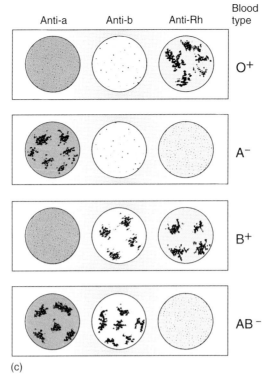

Anti-a	Anti-b	Anti-Rh	Blood type
			O⁺

Interpretation of blood typing
Figure 17.10

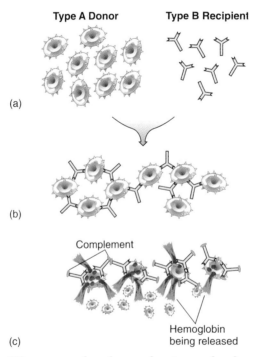

Type A Donor   **Type B Recipient**

(a)

(b)

Complement

(c)   Hemoglobin
                   being released

Microscopic view of a transfusion reaction
Figure 17.11

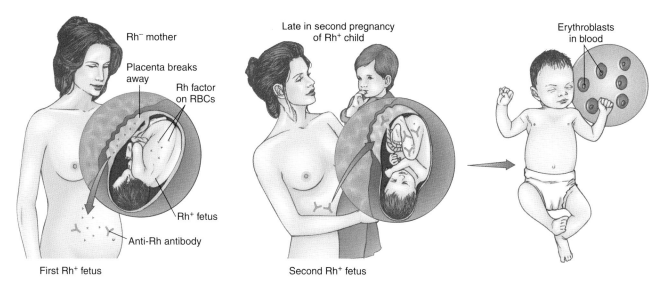

(a) The development and aftermath of Rh sensitization
Initial sensitization of the maternal immune system to fetal Rh$^+$ factor occurs when fetal cells leak into the Rh$^-$ mother's circulation late in pregnancy, or during delivery, when the placenta tears away. The child will escape hemolytic disease in most instances, but the mother, now sensitized, will be capable of an immediate reaction to a second Rh$^+$ fetus and its Rh-factor antigen. At that time, the mother's anti-Rh antibodies pass into the fetal circulation and elicit severe hemolysis in the fetus and neonate.

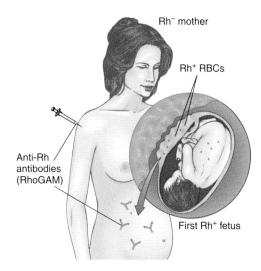

(b) Prevention of erythroblastosis fetalis with anti-Rh immune globulin (RhoGAM)
Injecting a mother who is at risk with RhoGAM during her first Rh$^+$ pregnancy helps to inactivate and remove the fetal Rh-positive cells before her immune system can react and develop sensitivity.

Development and control of Rh incompatibility
Figure 17.12

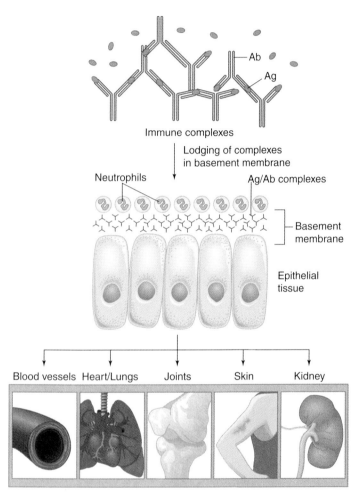

Immune complexes

Lodging of complexes
in basement membrane

Neutrophils

Ag/Ab complexes

Basement
membrane

Epithelial
tissue

Blood vessels   Heart/Lungs   Joints   Skin   Kidney

Major organs that can be targets
of immune complex deposition

Steps:
1. Antibody combines with excess soluble antigen,
   forming large quantities of Ab/Ag complexes.

2. Circulating immune complexes become lodged
   in the basement membrane of epithelia in sites
   such as kidney, lungs, joints, skin.

3. Fragments of complement cause release of histamine
   and other mediator substances.
4. Neutrophils migrate to the site of immune complex
   deposition and release enzymes that cause severe
   damage in the tissues and organs involved.

Pathogenesis of immune complex disease
Figure 17.13

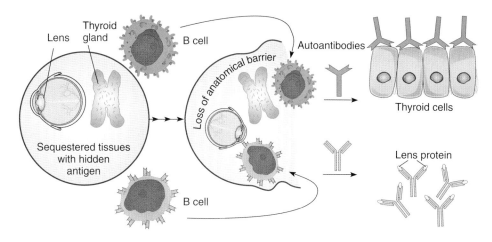

(a) **Sequestered Antigen Theory**

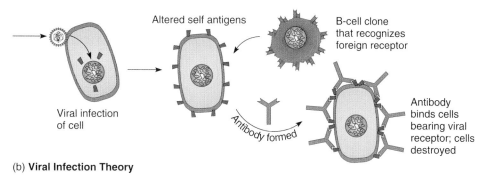

(b) **Viral Infection Theory**

Possible explanations for autoimmunity
Figure 17.14

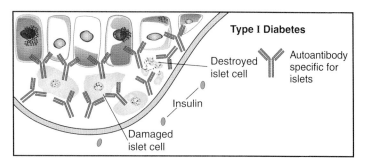

The autoimmune component in diabetes mellitus, type I
Figure 17.16

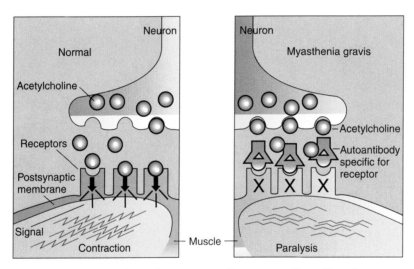

Mechanism for involvement of autoantibodies in myasthenia gravis
Figure 17.17

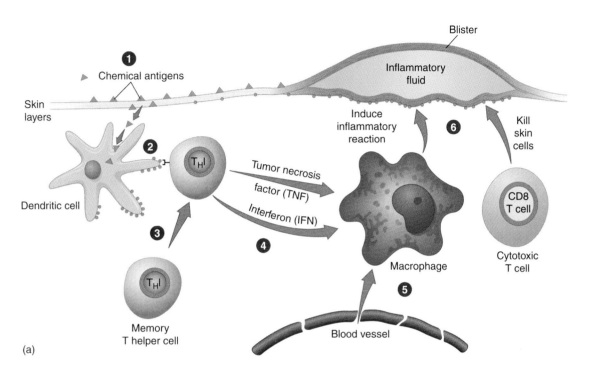

Chemical antigens

Skin layers

Dendritic cell

T$_H$I

Memory T helper cell

(a)

Tumor necrosis factor (TNF)

Interferon (IFN)

Induce inflammatory reaction

Macrophage

Blood vessel

Blister

Inflammatory fluid

Kill skin cells

CD8 T cell

Cytotoxic T cell

1 Lipid-soluble chemicals are absorbed by the skin.

2 Dendritic cells close to the epithelium pick up the allergen, process it, and display it on MHC receptors.

3 Previously sensitized T$_H$I cells recognize the presented allergen.

4 Sensitized T$_H$I cells are activated to secrete cytokines (IFN, TNF) that attract macrophages and cytotoxic T cells to the site. **5**

6 Macrophage releases mediators that stimulate a strong, local inflammatory reaction. Cytotoxic T cells directly kill cells and damage the skin. Fluid-filled blisters result.

Contact dermatitis
Figure 17.19

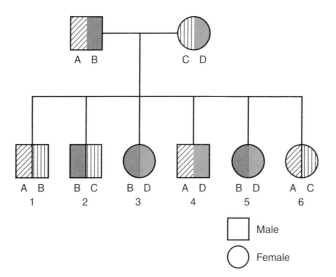

The pattern of inheritance of MHC (HLA) genes
Figure 17.20

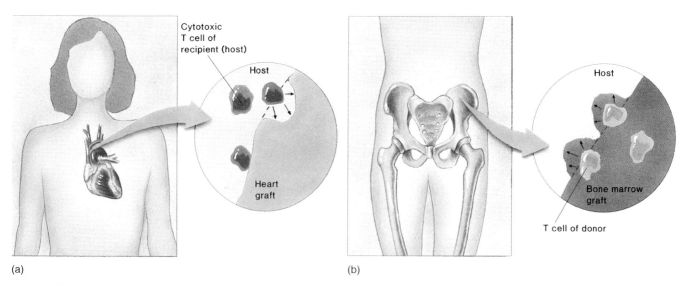

(a)

(b)

Potential reactions in transplantation
Figure 17.21

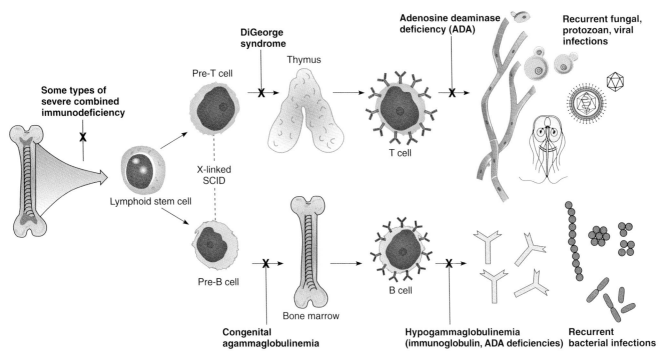

The stages of development and the functions of B cells and T cells, whose failure causes immunodeficiencies
Figure 17.22

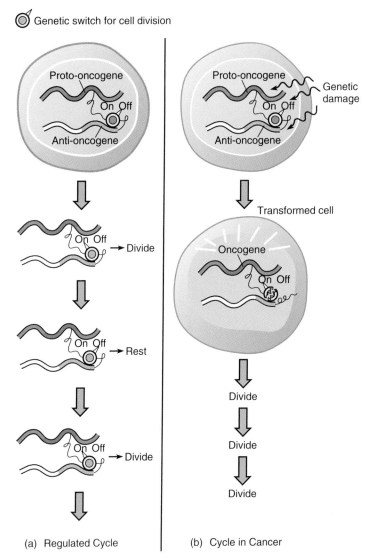

○ Genetic switch for cell division

Proto-oncogene
On Off
Anti-oncogene

Proto-oncogene
On Off
Anti-oncogene
Genetic damage

On Off → Divide

Transformed cell

On Off → Rest

Oncogene
On Off

On Off → Divide

Divide

Divide

Divide

(a) Regulated Cycle

(b) Cycle in Cancer

Common pathway for neoplasias
Figure 17.25

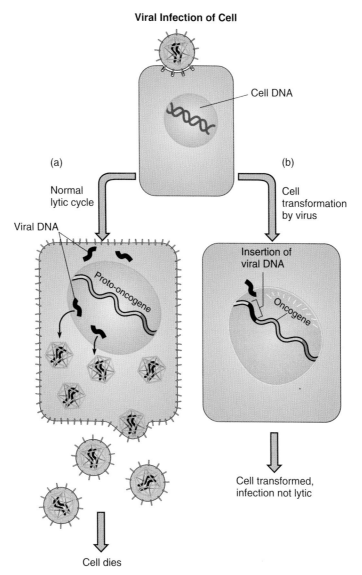

Viral Infection of Cell

Cell DNA

(a)

Normal
lytic cycle

Viral DNA

Proto-oncogene

Cell dies

(b)

Cell
transformation
by virus

Insertion of
viral DNA

Oncogene

Cell transformed,
infection not lytic

Outcomes of viral infection
Figure 17.26

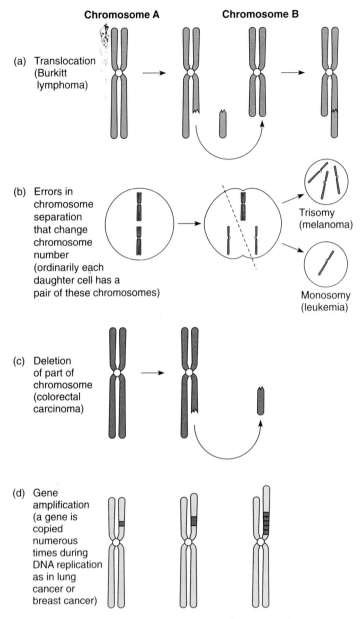

Chromosome A   **Chromosome B**

(a) Translocation
(Burkitt
lymphoma)

(b) Errors in
chromosome
separation
that change
chromosome
number
(ordinarily each
daughter cell has a
pair of these chromosomes)

Trisomy
(melanoma)

Monosomy
(leukemia)

(c) Deletion
of part of
chromosome
(colorectal
carcinoma)

(d) Gene
amplification
(a gene is
copied
numerous
times during
DNA replication
as in lung
cancer or
breast cancer)

**Alterations in chromosomes that can be
implicated in some cancers (a-d)**
Figure 17.27

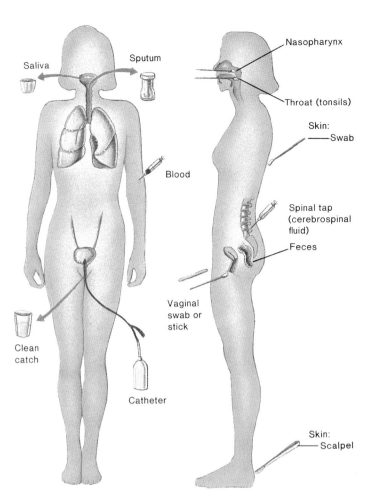

Sampling sites and methods of collection for clinical laboratories
Figure A

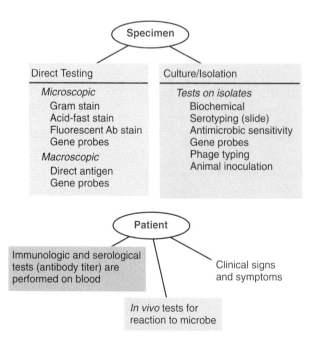

A scheme of specimen isolation and identification
Figure B

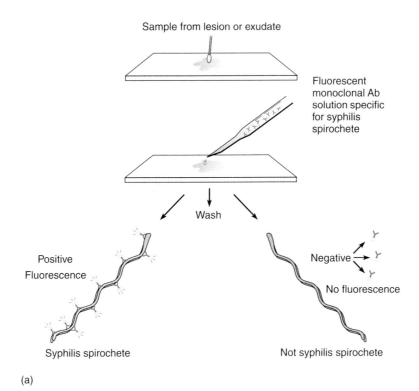

Sample from lesion or exudate

Fluorescent monoclonal Ab solution specific for syphilis spirochete

Wash

Positive Fluorescence

Syphilis spirochete

Negative

No fluorescence

Not syphilis spirochete

(a)

Direct fluorescent antigen test results for *Treponema pallidum,* **the syphilis spirochete, and an unrelated spirochete**
Figure D

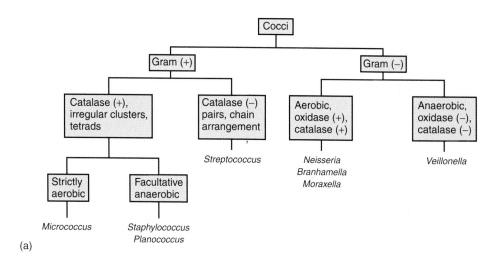

(a)

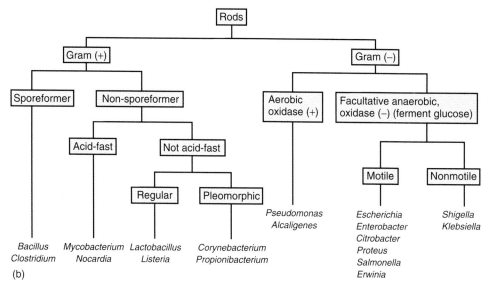

(b)

Flowchart to separate primary genera of gram-positive and gram-negative bacteria
Figure E

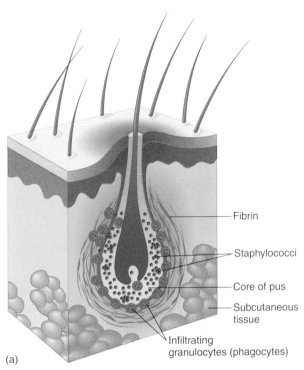

Fibrin

Staphylococci

Core of pus

Subcutaneous
tissue

Infiltrating
granulocytes (phagocytes)

(a)

**Cutaneous lesions of *Staphylococcus
aureus***
Figure 18.3

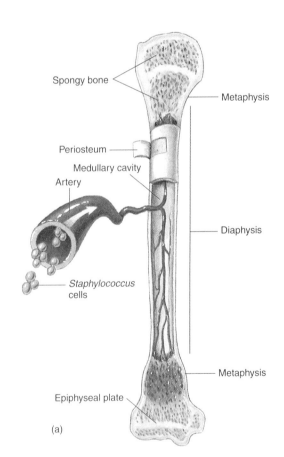

Spongy bone

Metaphysis

Periosteum

Medullary cavity

Artery

Staphylococcus
cells

Diaphysis

Metaphysis

Epiphyseal plate

(a)

**Staphylococcal osteomyelitis in a
long bone**
Figure 18.4

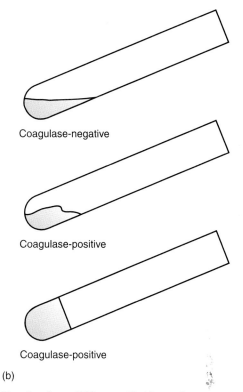

Coagulase-negative

Coagulase-positive

Coagulase-positive

(b)

Tests for differentiating the genus *Staphylococcus* from *Streptococcus* and for identifying *Staphylococcus aureus*
Figure 18.6

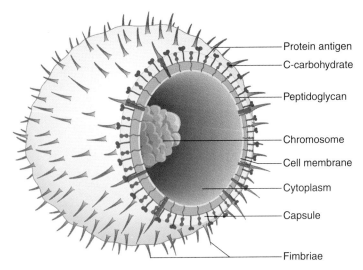

Cutaway view of group A *Streptococcus*
Figure 18.10

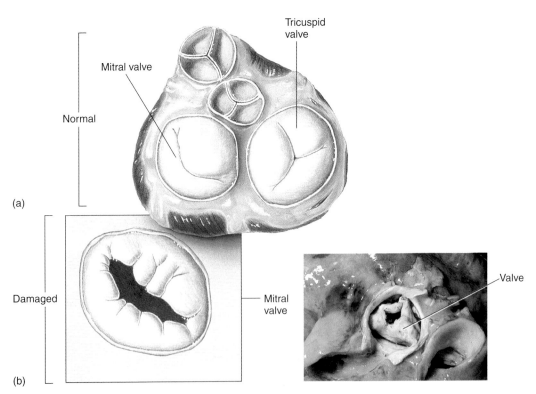

The cardiac complications of rheumatic fever
Figure 18.13
b2: CDC/Dr. Edwin P. Ewing, Jr.

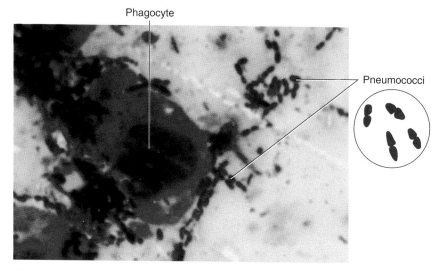

Diagnosing *Streptococcus pneumoniae*
Figure 18.17
© A.M. Siegelman/Visuals Unlimited

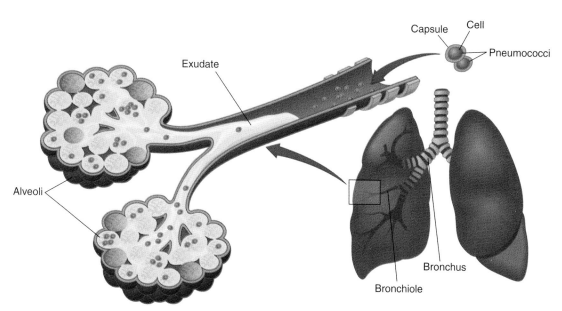

The course of bacterial pneumonia
Figure 18.18

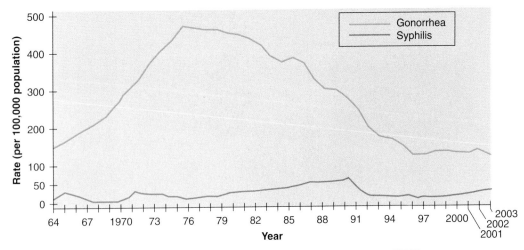

Gonorrhea and Syphilis—Reported Rates: United States, 1964–2003

Comparative incidence of two reportable infectious STDs
Figure 18.22

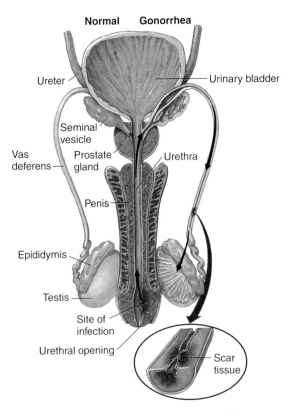

Gonorrheal damage to the male reproductive tract
Figure 18.23

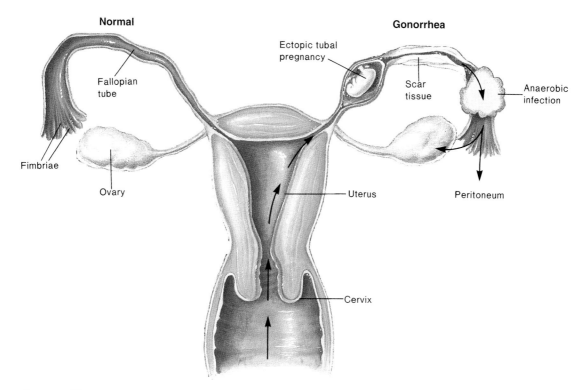

Ascending gonorrhea in women
Figure 18.24

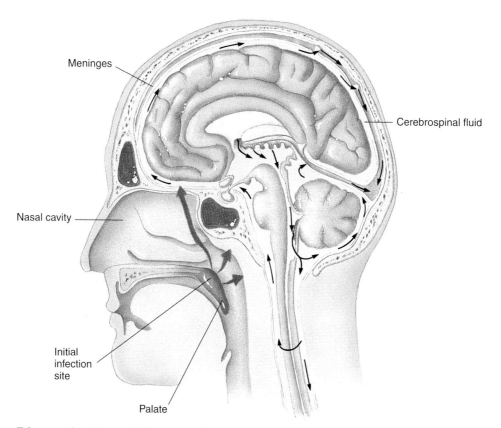

Dissemination of the meningococcus from a nasopharyngeal infection
Figure 18.27

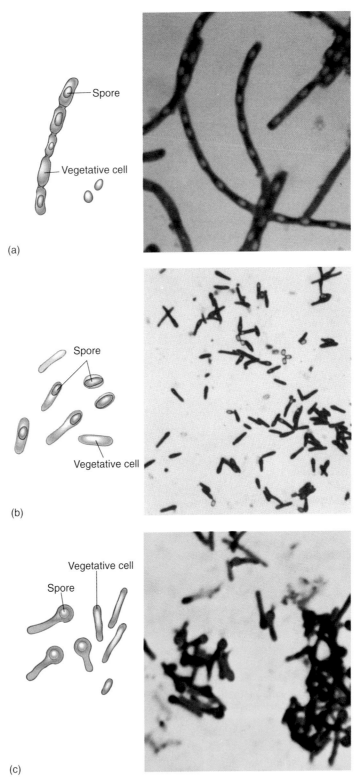

Spore

Vegetative cell

(a)

Spore

Vegetative cell

(b)

Vegetative cell

Spore

(c)

Examples of endospore-forming pathogens
Figure 19.1

a: © A.M. Siegelman/Visuals Unlimited; **b:** © George J. Wilder/ Visuals Unlimited; **c:** © John D. Cunningham/Visuals Unlimited

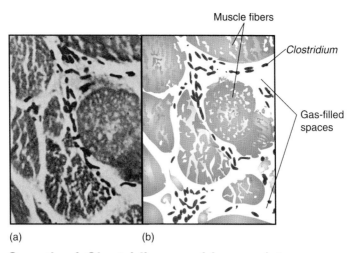

Growth of *Clostridium perfringens* (plump rods), causing gas formation and separation of the fibers

Figure 19.3

a: From M.A. Boyd et al, *Journal of Medical Microbiology,* 5:459, 1972. Reprinted by permission of Longarm Group, Ltd.
© Pathological Society of Great Britain and Ireland

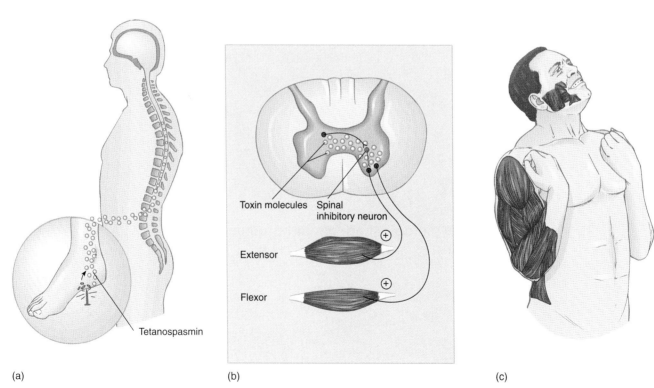

The events in tetanus

Figure 19.7

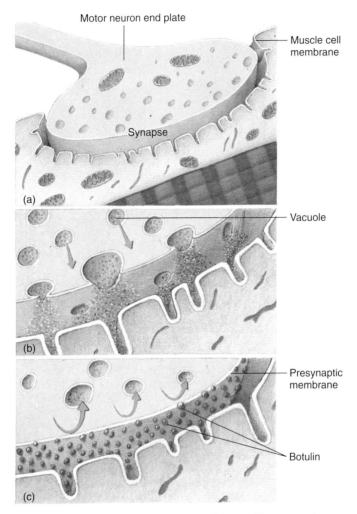

**The physiological effects of botulism toxin
(botulin)**
Figure 19.9

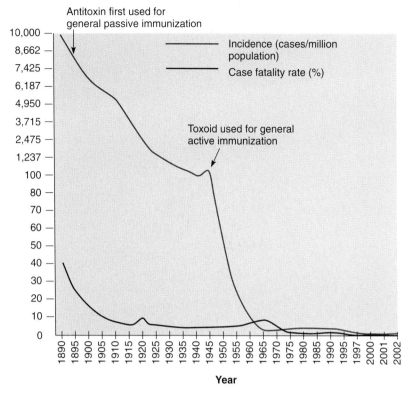

The incidence and case fatality rates for diphtheria in the United States during the last 110 years
Figure 19.12

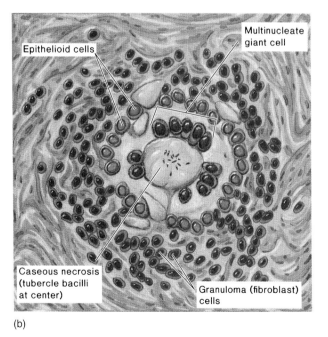

(b)

Tubercle formation
Figure 19.16

267

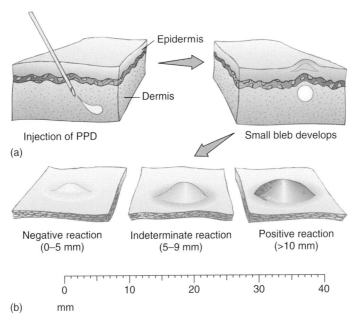

(a)

Epidermis

Dermis

Injection of PPD

Small bleb develops

Negative reaction
(0–5 mm)

Indeterminate reaction
(5–9 mm)

Positive reaction
(>10 mm)

0   10   20   30   40

(b)   mm

Skin testing for tuberculosis
Figure 19.17

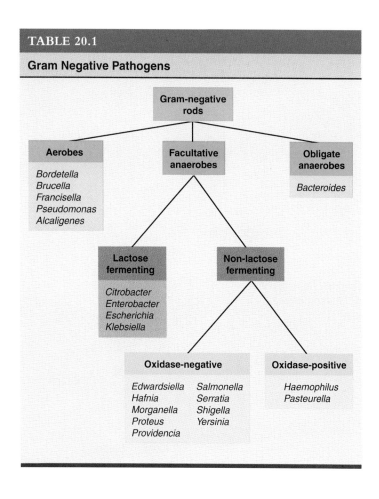

TABLE 20.1

Gram Negative Pathogens

Gram-negative rods

Aerobes

Bordetella
Brucella
Francisella
Pseudomonas
Alcaligenes

Facultative anaerobes

Obligate anaerobes

Bacteroides

Lactose fermenting

Citrobacter
Enterobacter
Escherichia
Klebsiella

Non-lactose fermenting

Oxidase-negative

Edwardsiella	Salmonella
Hafnia	Serratia
Morganella	Shigella
Proteus	Yersinia
Providencia	

Oxidase-positive

Haemophilus
Pasteurella

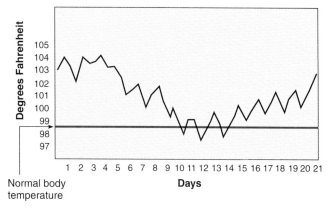

Normal body temperature

The temperature cycle in classic brucellosis
Figure 20.4

Source: Data from A. Smith, *Principles of Microbiology,* 10th ed., 1985.

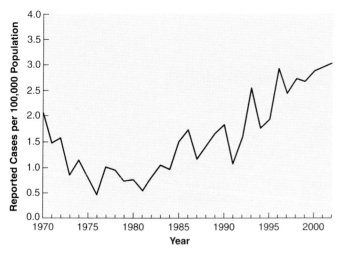

Prevalence of pertussis (whooping cough) in the United States

Figure 20.5

Source: Centers for Disease Control and Prevention. Summary of notifiable diseases, United States, 2002.

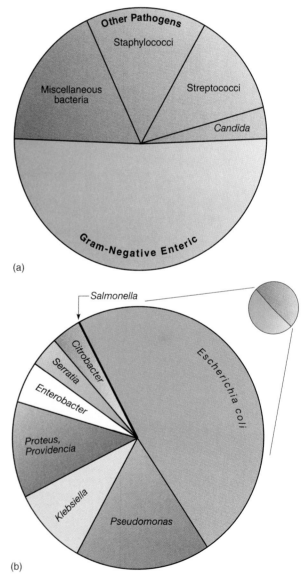

(a)

(b)

Bacteria that account for the majority of nosocomial infections

Figure 20.8

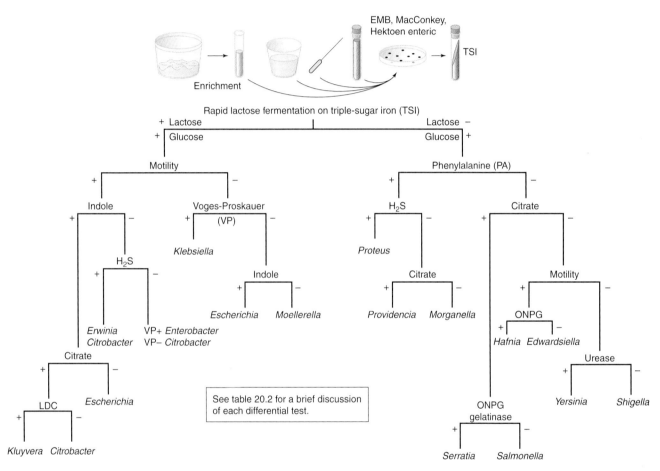

Procedures for isolating and identifying selected enteric genera
Figure 20.9

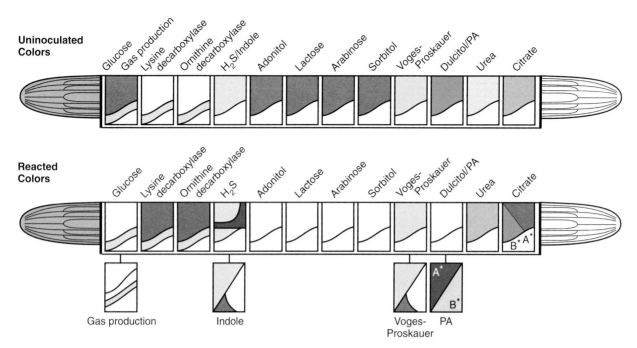

BBL Enterotube™ II (Roche Diagnostics), a miniaturized, multichambered tube used for rapid biochemical testing of enterics

Figure 20.11

Reprinted courtesy of Becton, Dickinson and Company

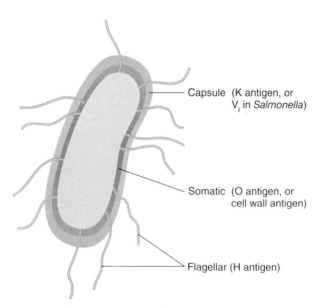

Antigenic structures in gram-negative enteric rods

Figure 20.12

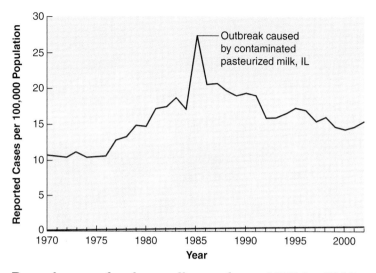

Prevalence of salmonelloses from 1970 to 2002
Figure 20.16

Source: Centers for Disease Control and Prevention. Summary of notifiable diseases, United States, 2002.

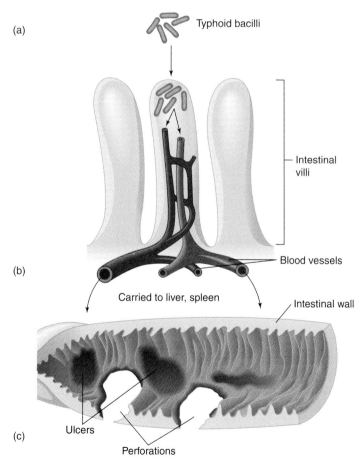

The phases of typhoid fever
Figure 20.17

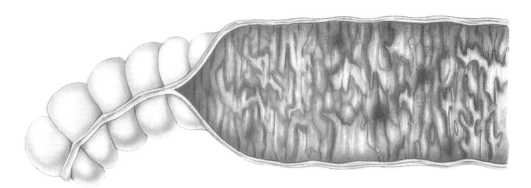

The appearance of the large intestinal mucosa in *Shigella* (bacillary) dysentery
Figure 20.18

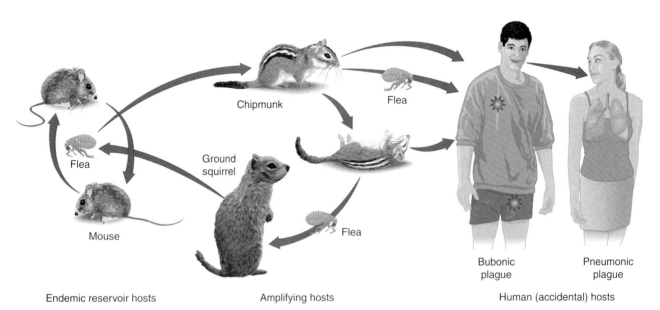

Chipmunk

Flea

Flea

Ground squirrel

Mouse

Flea

Bubonic plague

Pneumonic plague

Endemic reservoir hosts   Amplifying hosts   Human (accidental) hosts

The infection cycle of *Yersinia pestis* simplified for clarity
Figure 20.20

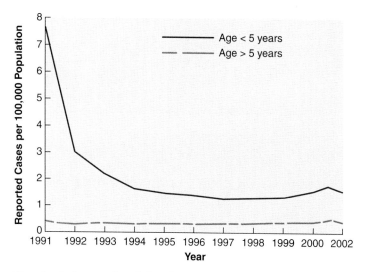

Bacterial meningitis due to *Haemophilus influenzae* infection
Figure 20.23

(a)

Typical spirochete
Figure 21.1

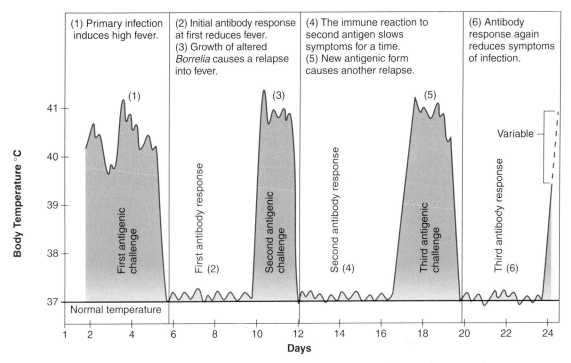

The pattern in relapsing fever, based on symptoms (fever) over time
Figure 21.10

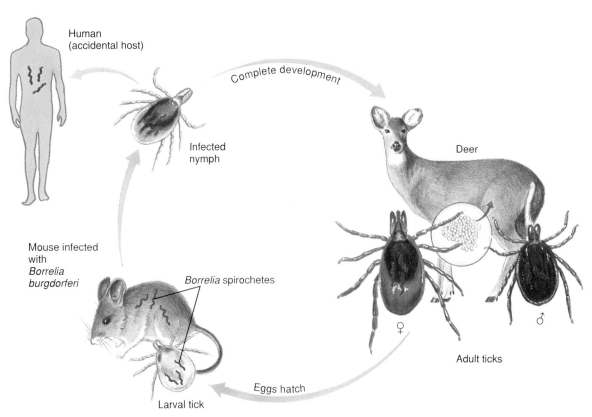

The cycle of Lyme disease in the northeastern United States
Figure 21.11

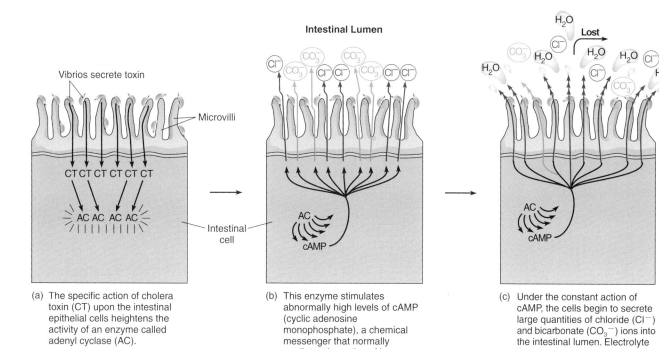

Intestinal Lumen

(a) The specific action of cholera toxin (CT) upon the intestinal epithelial cells heightens the activity of an enzyme called adenyl cyclase (AC).

(b) This enzyme stimulates abnormally high levels of cAMP (cyclic adenosine monophosphate), a chemical messenger that normally mediates the action of hormones on cells, but in higher concentrations promotes removal of anions (chloride and carbonate) by the cell membrane.

(c) Under the constant action of cAMP, the cells begin to secrete large quantities of chloride (Cl^-) and bicarbonate (CO_3^-) ions into the intestinal lumen. Electrolyte loss is followed by water loss from epithelial cells, which is what causes the major symptoms.

Pathogenesis of cholera
Figure 21.14

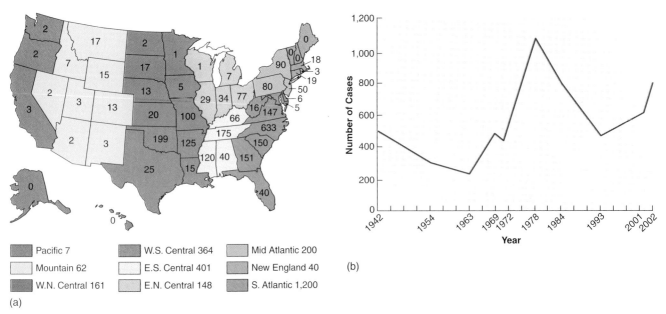

Trends in infection for Rocky Mountain spotted fever
Figure 21.18

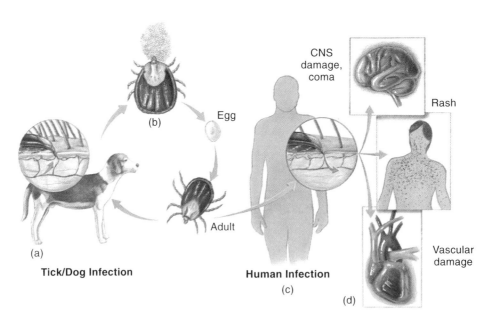

The transmission cycle in Rocky Mountain spotted fever
Figure 21.19

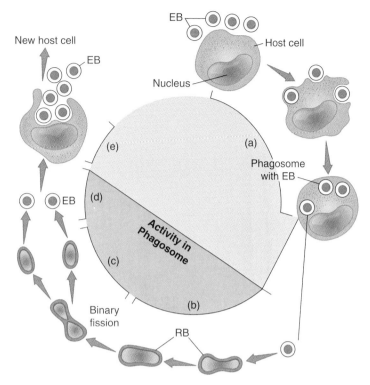

The life cycle of *Chlamydia*
Figure 21.23

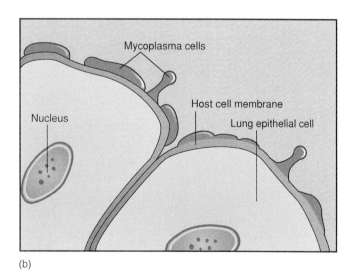

(b)

The morphology of mycoplasmas
Figure 21.27

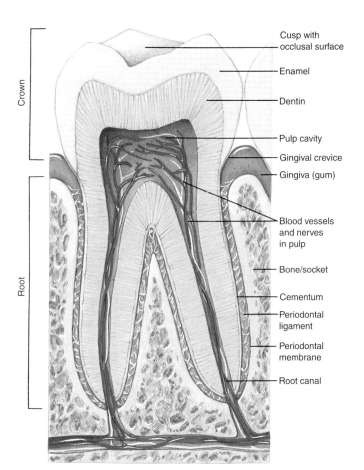

The anatomy of a tooth
Figure 21.28

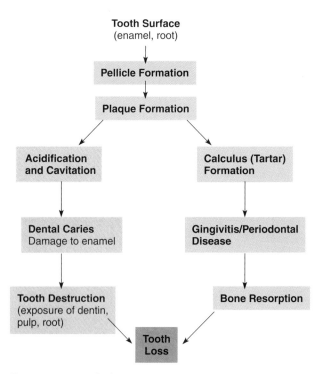

Summary of the events leading to dental caries, periodontal disease, and bone and tooth loss
Figure 21.29

(a)

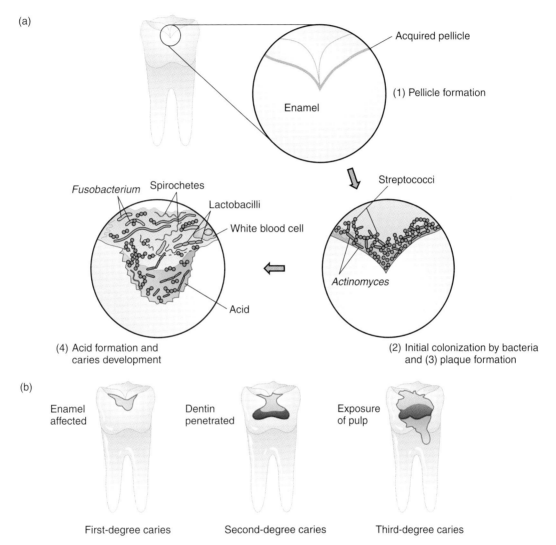

Acquired pellicle

(1) Pellicle formation

Enamel

Streptococci

Actinomyces

(2) Initial colonization by bacteria
and (3) plaque formation

Fusobacterium   Spirochetes

Lactobacilli

White blood cell

Acid

(4) Acid formation and
caries development

(b)

Enamel
affected

Dentin
penetrated

Exposure
of pulp

First-degree caries

Second-degree caries

Third-degree caries

Stages in plaque development and cariogenesis
Figure 21.30

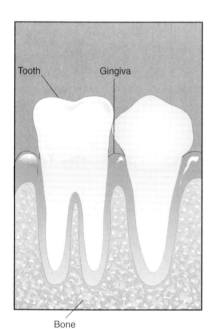

(a) Normal, nondiseased state of tooth, gingiva, and bone.

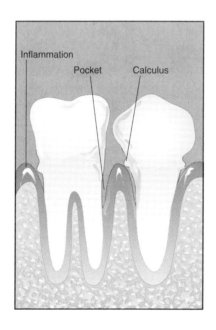

(b) Calculus buildup and early gingivitis.

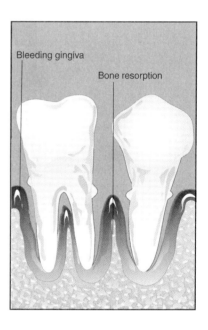

(c) Late-stage periodontitis, with tissue destruction, deep pocket formation, loosening of teeth, and bone loss.

Stages in soft-tissue infection, gingivitis, and periodontitis
Figure 21.33

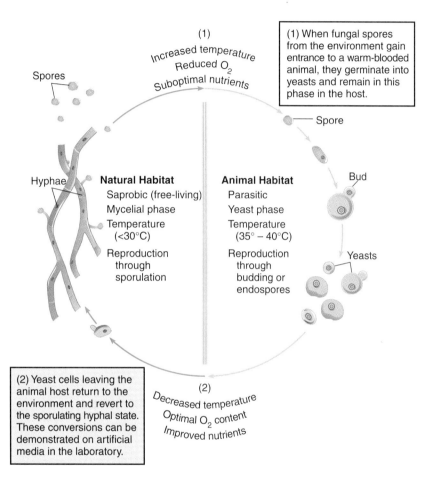

(1)
Increased temperature
Reduced O$_2$
Suboptimal nutrients

Spores

(1) When fungal spores from the environment gain entrance to a warm-blooded animal, they germinate into yeasts and remain in this phase in the host.

Spore

Hyphae

Bud

Natural Habitat
Saprobic (free-living)
Mycelial phase
Temperature (<30°C)
Reproduction through sporulation

Animal Habitat
Parasitic
Yeast phase
Temperature (35° – 40°C)
Reproduction through budding or endospores

Yeasts

(2) Yeast cells leaving the animal host return to the environment and revert to the sporulating hyphal state. These conversions can be demonstrated on artificial media in the laboratory.

(2)
Decreased temperature
Optimal O$_2$ content
Improved nutrients

The general changes associated with thermal dimorphism, using generic hyphae, spores, and yeasts as examples
Figure 22.1

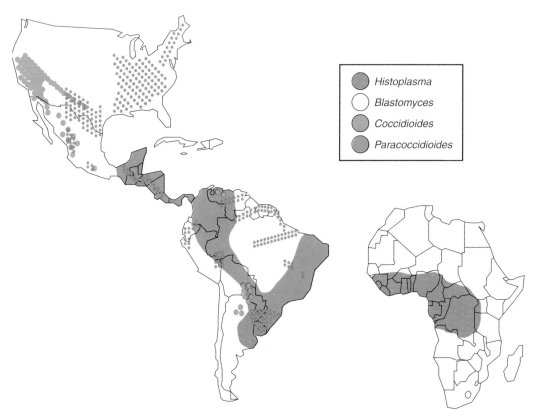

Distribution of the four fungal pathogens
Figure 22.2

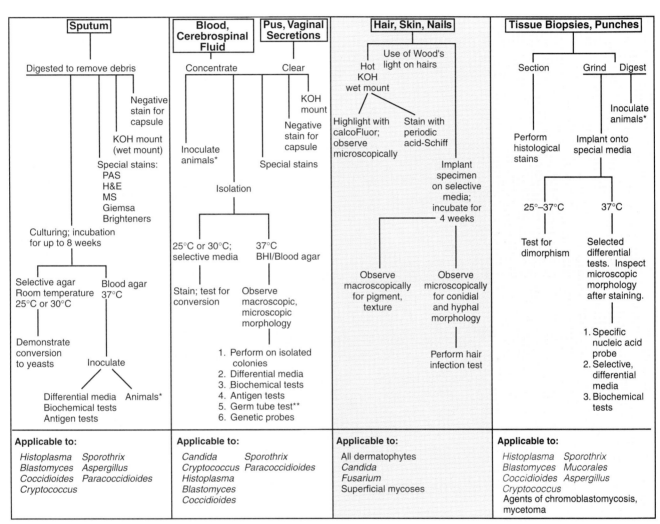

Sputum	Blood, Cerebrospinal Fluid	Pus, Vaginal Secretions	Hair, Skin, Nails	Tissue Biopsies, Punches

Digested to remove debris

Negative stain for capsule

KOH mount (wet mount)

Special stains:
PAS
H&E
MS
Giemsa
Brighteners

Culturing; incubation for up to 8 weeks

Selective agar
Room temperature
25°C or 30°C

Blood agar
37°C

Demonstrate conversion to yeasts

Inoculate

Differential media   Animals*
Biochemical tests
Antigen tests

Concentrate   Clear

KOH mount

Negative stain for capsule

Inoculate animals*

Special stains

Isolation

25°C or 30°C;   37°C
selective media   BHI/Blood agar

Stain; test for   Observe
conversion   macroscopic,
                         microscopic
                         morphology

1. Perform on isolated colonies
2. Differential media
3. Biochemical tests
4. Antigen tests
5. Germ tube test**
6. Genetic probes

Use of Wood's light on hairs

Hot
KOH
wet mount

Highlight with   Stain with
calcoFluor;   periodic
observe   acid-Schiff
microscopically

Implant specimen on selective media; incubate for 4 weeks

Observe   Observe
macroscopically   microscopically
for pigment,   for conidial
texture   and hyphal
                        morphology

Perform hair infection test

Section   Grind   Digest

Inoculate animals*

Perform   Implant onto
histological   special media
stains

25°–37°C   37°C

Test for   Selected
dimorphism   differential
                 tests. Inspect
                 microscopic
                 morphology
                 after staining.

1. Specific nucleic acid probe
2. Selective, differential media
3. Biochemical tests

Applicable to:

Histoplasma	Sporothrix
Blastomyces	Aspergillus
Coccidioides	Paracoccidioides
Cryptococcus	

Applicable to:

Candida	Sporothrix
Cryptococcus	Paracoccidioides
Histoplasma	
Blastomyces	
Coccidioides	

Applicable to:

All dermatophytes
Candida
Fusarium
Superficial mycoses

Applicable to:

Histoplasma	Sporothrix
Blastomyces	Mucorales
Coccidioides	Aspergillus
Cryptococcus	

Agents of chromoblastomycosis, mycetoma

*Animal inoculation is performed only to help diagnose systemic mycoses when other methods are unavailable or indeterminant.
**Some yeasts, when incubated in serum for 2 to 4 hours, sprout tiny hyphal tubes called germ tubes. *Candida albicans* is identified by this characteristic.

Methods of processing specimens in fungal disease
Figure 22.3

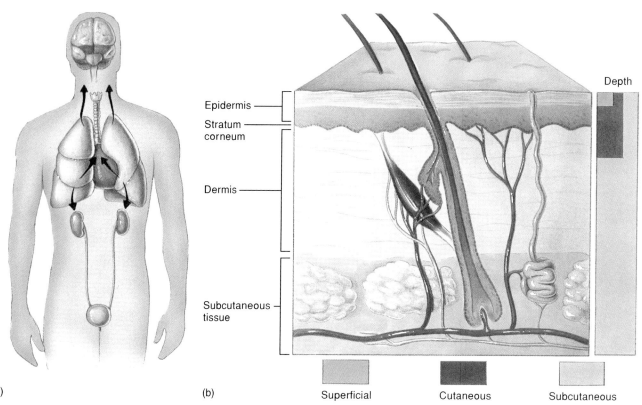

(a)

(b)

Epidermis
Stratum
corneum

Dermis

Subcutaneous
tissue

Depth

Superficial   Cutaneous   Subcutaneous

Levels of invasion by fungal pathogens
Figure 22.5

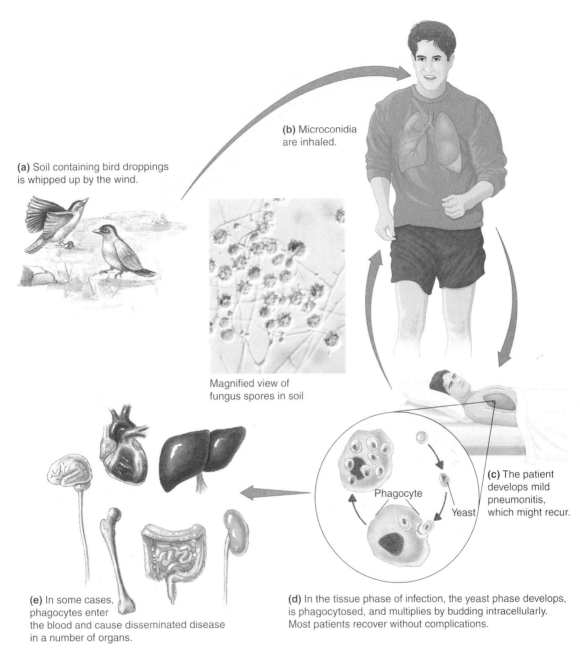

(a) Soil containing bird droppings is whipped up by the wind.

(b) Microconidia are inhaled.

Magnified view of fungus spores in soil

Phagocyte

Yeast

(c) The patient develops mild pneumonitis, which might recur.

(e) In some cases, phagocytes enter the blood and cause disseminated disease in a number of organs.

(d) In the tissue phase of infection, the yeast phase develops, is phagocytosed, and multiplies by budding intracellularly. Most patients recover without complications.

Events in *Histoplasma* infection and histoplasmosis
Figure 22.7

b: © A.M. Siegelman/Visuals Unlimited

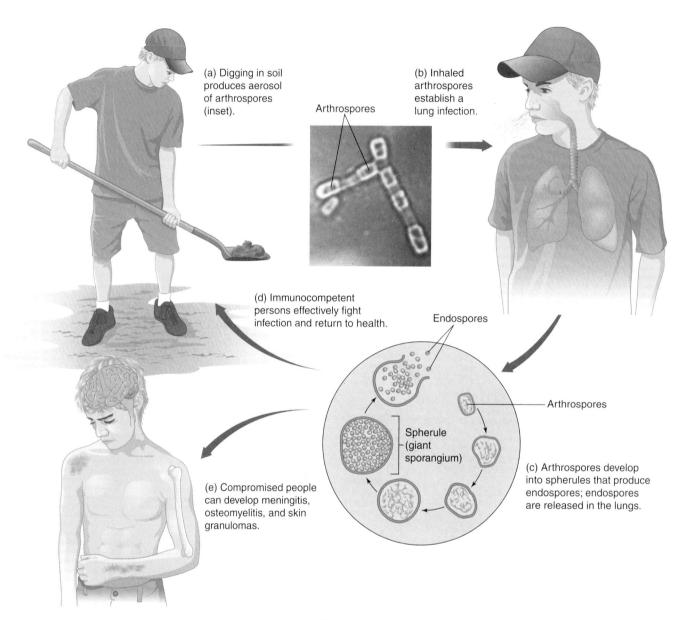

(a) Digging in soil produces aerosol of arthrospores (inset).

Arthrospores

(b) Inhaled arthrospores establish a lung infection.

(d) Immunocompetent persons effectively fight infection and return to health.

Endospores

Arthrospores

Spherule (giant sporangium)

(c) Arthrospores develop into spherules that produce endospores; endospores are released in the lungs.

(e) Compromised people can develop meningitis, osteomyelitis, and skin granulomas.

Events in *Coccidioides* infection and coccidioidomycosis
Figure 22.8

d: © Science VU-Charles Sutton/Visuals Unlimited

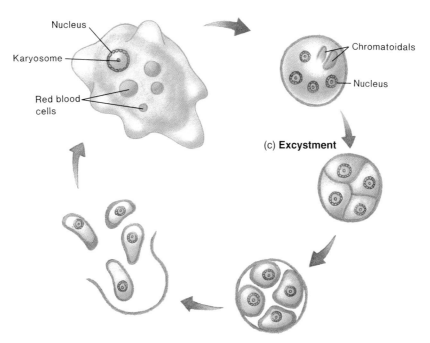

(a) **Trophozoite**

Nucleus

Karyosome

Red blood cells

(b) **Mature Cyst**

Chromatoidals

Nucleus

(c) **Excystment**

Cellular forms of *Entamoeba histolytica*
Figure 23.1

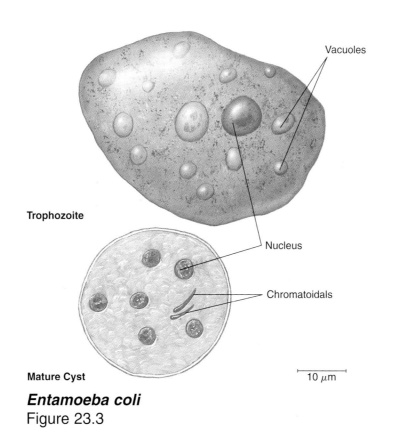

Vacuoles

Trophozoite

Nucleus

Chromatoidals

Mature Cyst

10 μm

Entamoeba coli
Figure 23.3

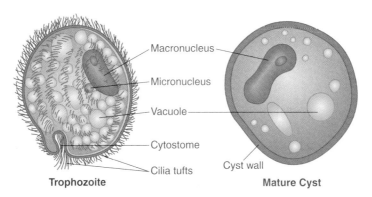

**The morphological anatomy of a trophozoite
and a mature cyst of *Balantidium coli***
Figure 23.5

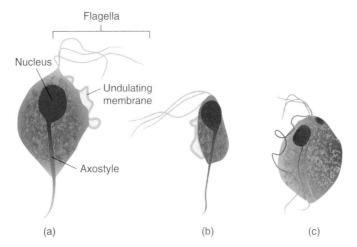

The trichomonads of humans
Figure 23.6

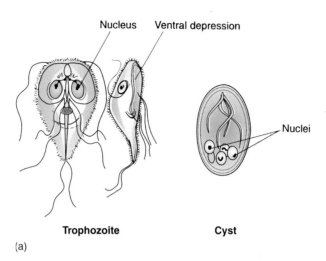

Nucleus   Ventral depression

Nuclei

Trophozoite   Cyst

(a)

Trophozoites and cysts of *Giardia lamblia*
Figure 23.7

TABLE 23.3

Cellular and Infective Stages of the Hemoflagellates

Genus/Species	Amastigote	Promastigote	Epimastigote	Trypomastigote
Leishmania	Intracellular in human macrophages	Found in sand fly gut; **infective to humans**	Does not occur	Does not occur
Trypanosoma brucei	Does not occur	Does not occur	Present in salivary gland of tsetse fly	In biting mouthparts of tsetse fly; **infective to humans**
Trypanosoma cruzi	Intracellular in human macrophages, liver, heart, spleen	Occurs	Present in gut of reduviid (kissing) bug	In feces of reduviid bug; **transferred to humans**

(a) The distribution of African trypanosomiasis.

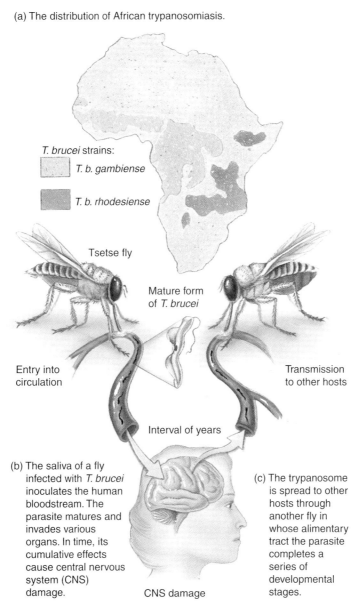

T. brucei strains:

T. b. gambiense

T. b. rhodesiense

Tsetse fly

Mature form
of T. brucei

Entry into
circulation

Transmission
to other hosts

Interval of years

(b) The saliva of a fly
infected with T. brucei
inoculates the human
bloodstream. The
parasite matures and
invades various
organs. In time, its
cumulative effects
cause central nervous
system (CNS)
damage.

CNS damage

(c) The trypanosome
is spread to other
hosts through
another fly in
whose alimentary
tract the parasite
completes a
series of
developmental
stages.

**The generalized cycle between humans and the
tsetse fly vector**
Figure 23.8

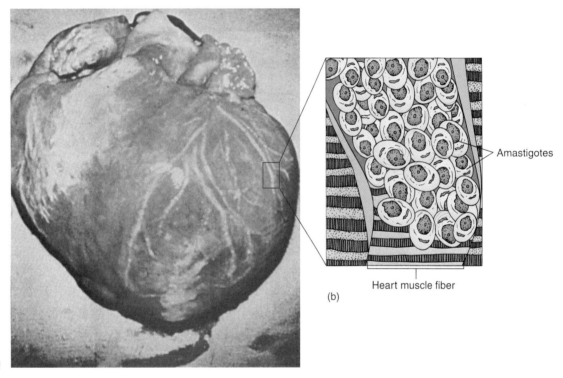

(a)

(b)

Amastigotes

Heart muscle fiber

Heart pathology in Chagas disease
Figure 23.9

a: Reprinted from Katz, Despommier, and Dwadz, "Parasitic Diseases," Springer-Verlag. Photo by T. Jones

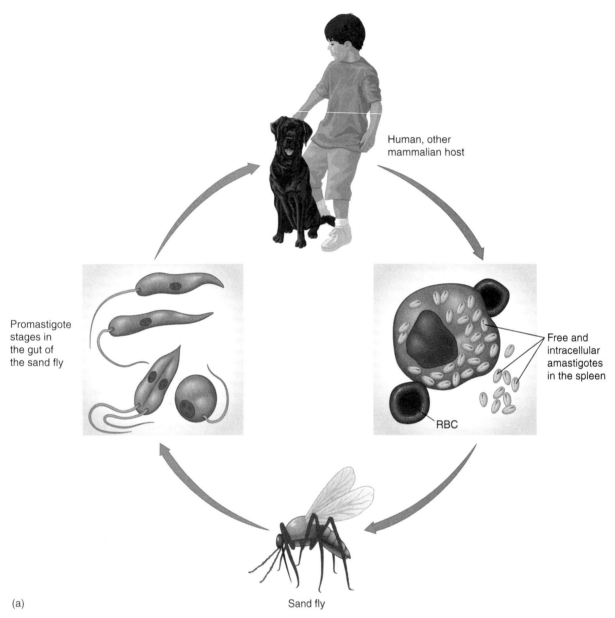

Human, other
mammalian host

Promastigote
stages in
the gut of
the sand fly

Free and
intracellular
amastigotes
in the spleen

RBC

(a)

Sand fly

The life cycle and pathology of *Leishmania* species
Figure 23.10

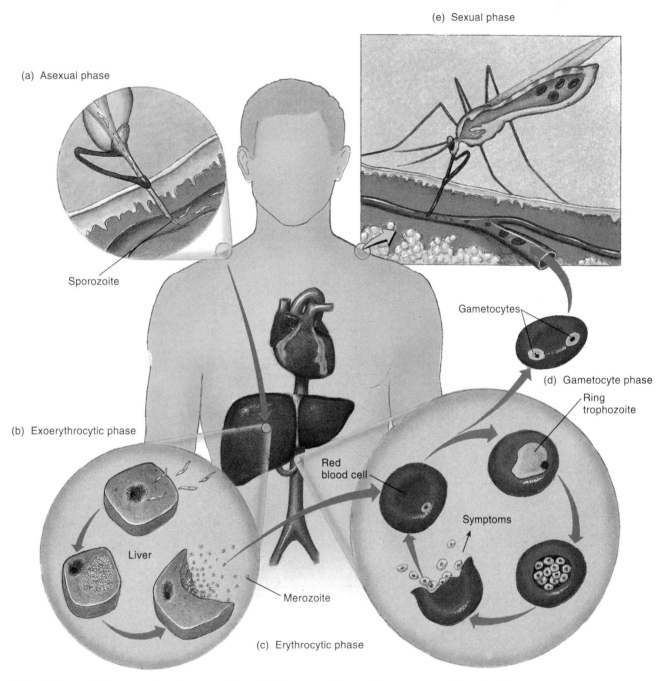

(e) Sexual phase

(a) Asexual phase

Sporozoite

Gametocytes

(d) Gametocyte phase

Ring trophozoite

(b) Exoerythrocytic phase

Red blood cell

Liver

Symptoms

Merozoite

(c) Erythrocytic phase

The life and transmission cycle of *Plasmodium,* the cause of malaria
Figure 23.11

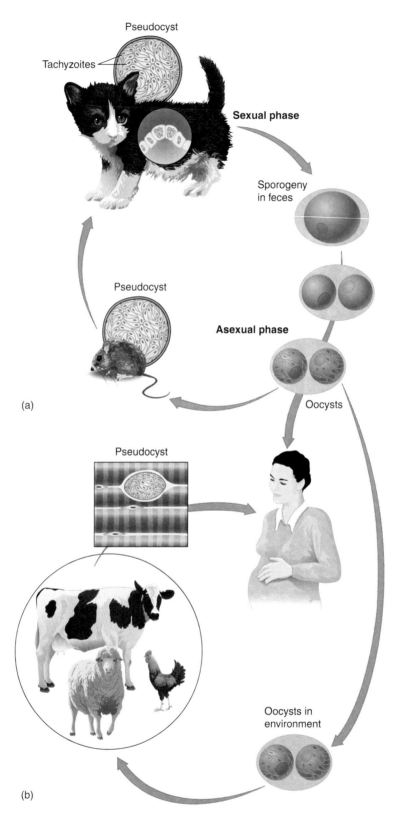

The life cycle and morphological forms of
Toxoplasma gondii
Figure 23.13

(a)

(b)

Pseudocyst

Tachyzoites

Sexual phase

Sporogeny
in feces

Pseudocyst

Asexual phase

Oocysts

Pseudocyst

Oocysts in
environment

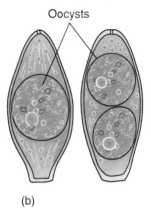

Oocysts

(b)

**Other apicomplexan
parasites**
Figure 23.15

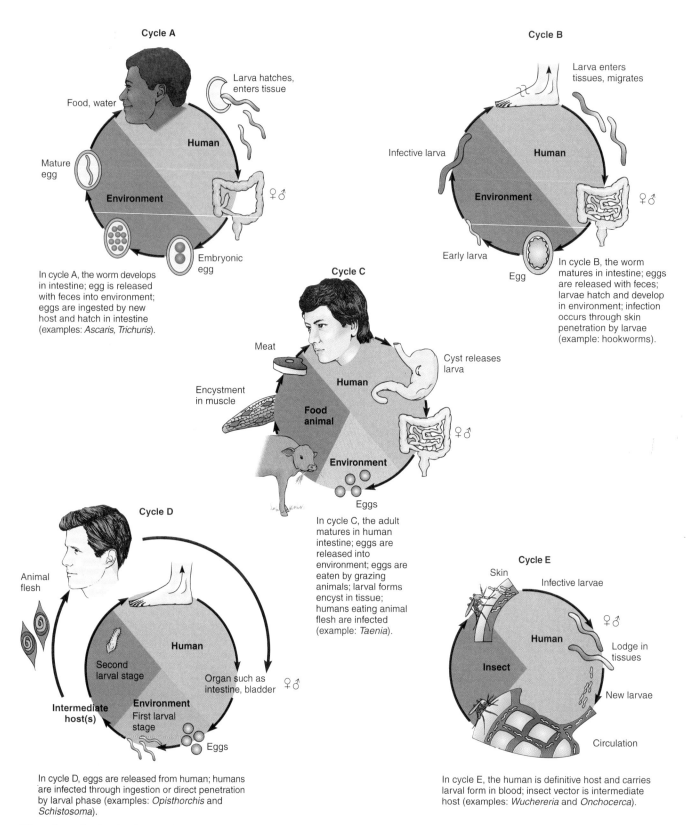

Cycle A

Food, water

Larva hatches, enters tissue

Human

Mature egg

Environment

♀♂

Embryonic egg

In cycle A, the worm develops in intestine; egg is released with feces into environment; eggs are ingested by new host and hatch in intestine (examples: *Ascaris, Trichuris*).

Cycle B

Larva enters tissues, migrates

Infective larva

Human

Environment

♀♂

Early larva

Egg

In cycle B, the worm matures in intestine; eggs are released with feces; larvae hatch and develop in environment; infection occurs through skin penetration by larvae (example: hookworms).

Cycle C

Meat

Cyst releases larva

Human

Encystment in muscle

Food animal

♀♂

Environment

Eggs

In cycle C, the adult matures in human intestine; eggs are released into environment; eggs are eaten by grazing animals; larval forms encyst in tissue; humans eating animal flesh are infected (example: *Taenia*).

Cycle D

Animal flesh

Human

Second larval stage

Organ such as intestine, bladder ♀♂

Intermediate host(s)

Environment
First larval stage

Eggs

In cycle D, eggs are released from human; humans are infected through ingestion or direct penetration by larval phase (examples: *Opisthorchis* and *Schistosoma*).

Cycle E

Skin

Infective larvae

Human

♀♂

Lodge in tissues

Insect

New larvae

Circulation

In cycle E, the human is definitive host and carries larval form in blood; insect vector is intermediate host (examples: *Wuchereria* and *Onchocerca*).

Five basic helminth life and transmission cycles
Figure 23.17

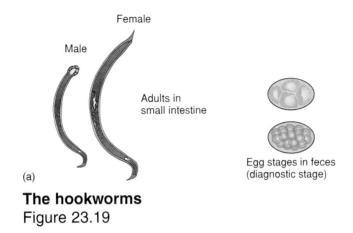

The hookworms
Figure 23.19

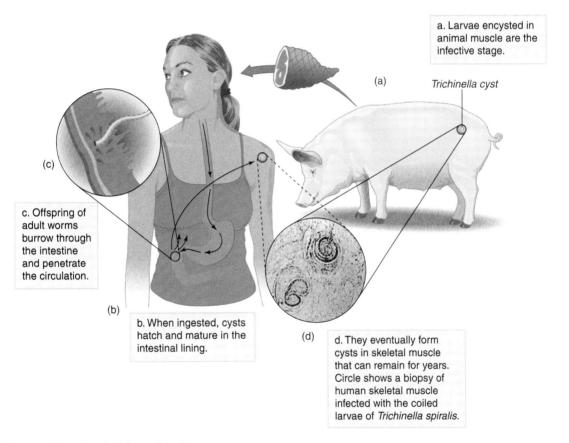

a. Larvae encysted in animal muscle are the infective stage.

Trichinella cyst

(a)

(c)

c. Offspring of adult worms burrow through the intestine and penetrate the circulation.

(b)

b. When ingested, cysts hatch and mature in the intestinal lining.

(d)

d. They eventually form cysts in skeletal muscle that can remain for years. Circle shows a biopsy of human skeletal muscle infected with the coiled larvae of *Trichinella spiralis*.

The cycle of trichinosis stage
Figure 23.21

(d inset) **Source:** Koneman et al., *Diagnostic Microbiology,* 4th ed., 1992. Reprinted by permission from J. B. Lippincott Co.

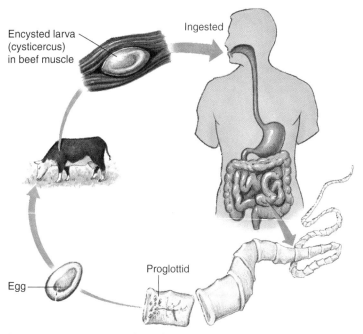

(c) A generalized diagram of the life cycle of the beef tapeworm *T. saginata.*

Tapeworm infestation in humans
Figure 23.25

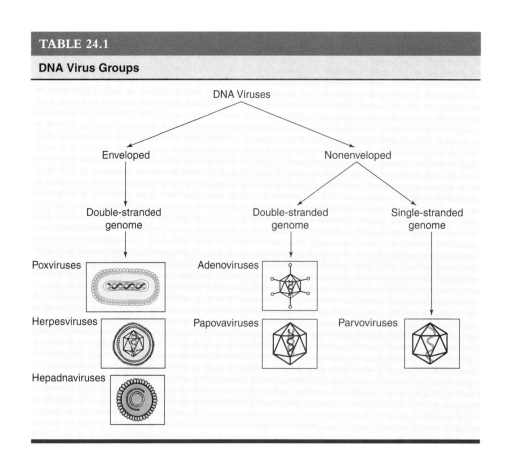

TABLE 24.1
DNA Virus Groups

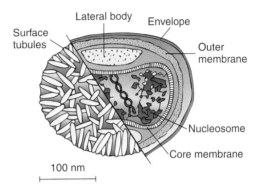

Poxviruses are larger and more complex than other viruses

Figure 24.1

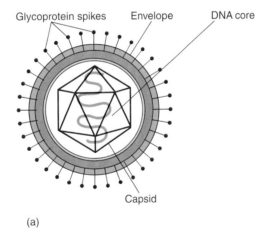

(a)

Herpesviruses

Figure 24.5

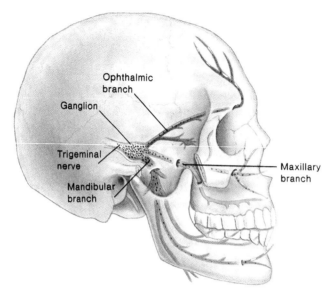

**The site of latency and routes of
recurrence in herpes simplex, type 1**
Figure 24.6

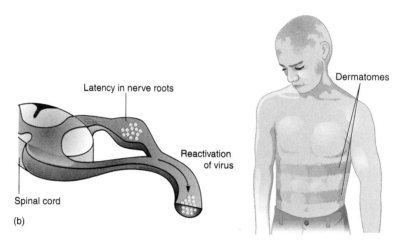

(b)

**The relationship between varicella (chickenpox)
and zoster (shingles) and the clinical appearance
of each**
Figure 24.12

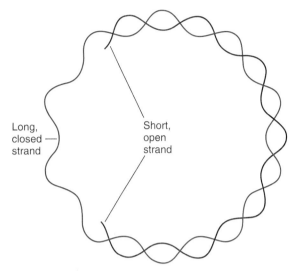

Long,
closed
strand

Short,
open
strand

**The nature of the DNA strand in
hepatitis B virus**
Figure 24.18

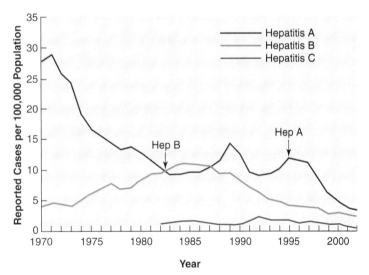

**The comparative incidence of viral hepatitis in
the United States, 1970–2002**
Figure 24.19

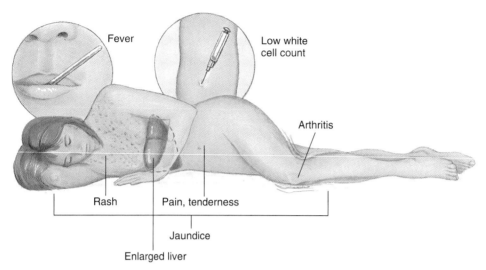

The clinical features of hepatitis B
Figure 24.21

TABLE 25.1

RNA Viruses

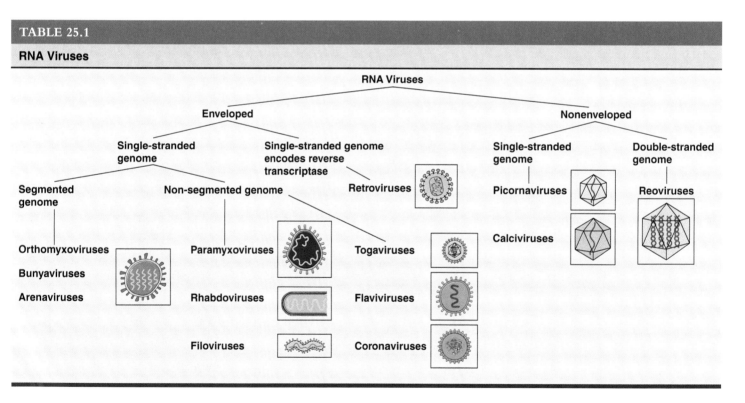

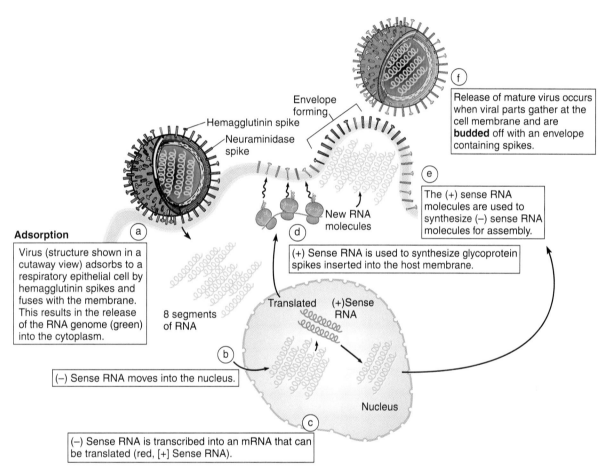

Adsorption

Virus (structure shown in a cutaway view) adsorbs to a respiratory epithelial cell by hemagglutinin spikes and fuses with the membrane. This results in the release of the RNA genome (green) into the cytoplasm.

Hemagglutinin spike

Neuraminidase spike

Envelope forming

Release of mature virus occurs when viral parts gather at the cell membrane and are **budded** off with an envelope containing spikes.

The (+) sense RNA molecules are used to synthesize (–) sense RNA molecules for assembly.

(+) Sense RNA is used to synthesize glycoprotein spikes inserted into the host membrane.

New RNA molecules

8 segments of RNA

Translated

(+)Sense RNA

(–) Sense RNA moves into the nucleus.

Nucleus

(–) Sense RNA is transcribed into an mRNA that can be translated (red, [+] Sense RNA).

Stages in cell invasion and disruption by the influenza virus
Figure 25.1

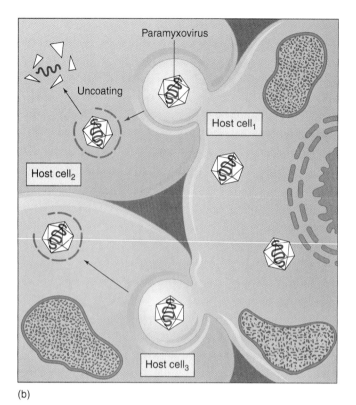

(b)

The effects of paramyxoviruses
Figure 25.2

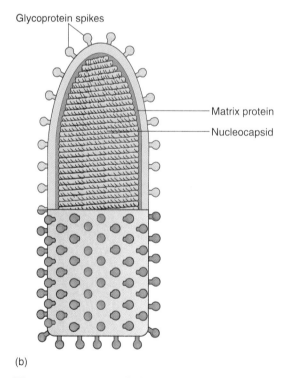

Glycoprotein spikes

Matrix protein

Nucleocapsid

(b)

The structure of the rabies virus
Figure 25.5

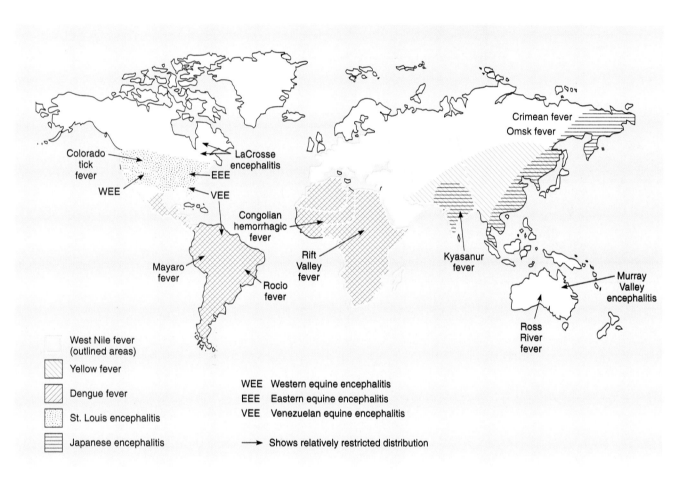

Worldwide distribution of major arboviral diseases
Figure 25.10

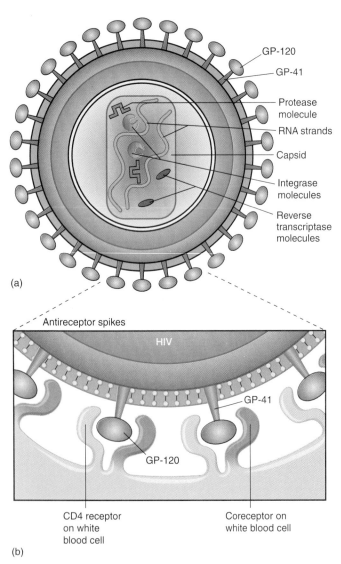

(a)

(b)

The general structure of HIV
Figure 25.12

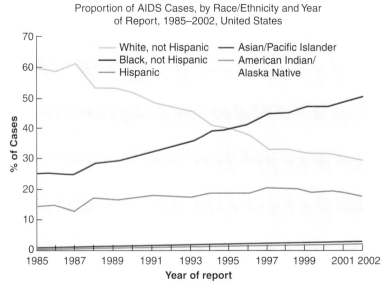

Proportion of AIDS Cases, by Race/Ethnicity and Year
of Report, 1985–2002, United States

White, not Hispanic — Asian/Pacific Islander
Black, not Hispanic — American Indian/
Hispanic — Alaska Native

% of Cases

Year of report

The changing face of AIDS
Figure 25.13

Source: Centers for Disease Control and Prevention

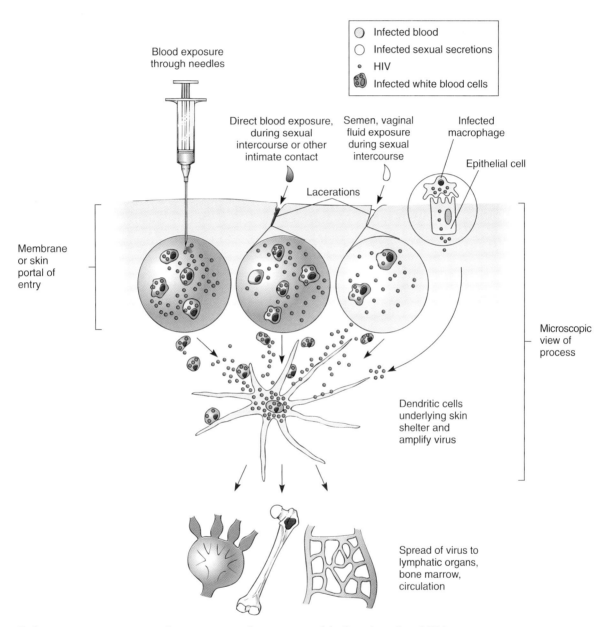

Primary sources and suggested routes of infection by HIV
Figure 25.14

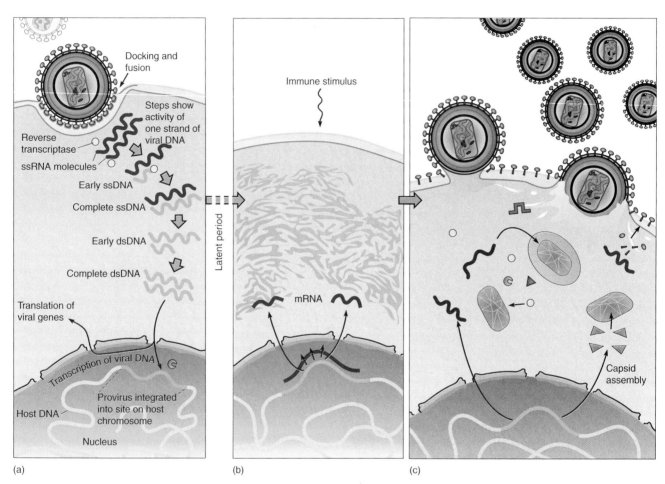

(a)

(b)

(c)

The virus is adsorbed and endocytosed, and the twin RNAs are uncoated. Reverse transcriptase catalyzes the synthesis of a single complementary strand of DNA (ssDNA). This single strand serves as a template for synthesis of a double strand (ds) of DNA. In latency, dsDNA is inserted into the host chromosome as a provirus.

After a latent period, various immune activators stimulate the infected cell, causing reactivation of the provirus genes and production of viral mRNA.

HIV mRNA is translated by the cell's synthetic machinery into virus components (capsid, reverse transcriptase, spikes), and the viruses are assembled. Budding of mature viruses lyses the infected cell.

The general multiplication cycle of HIV
Figure 25.15

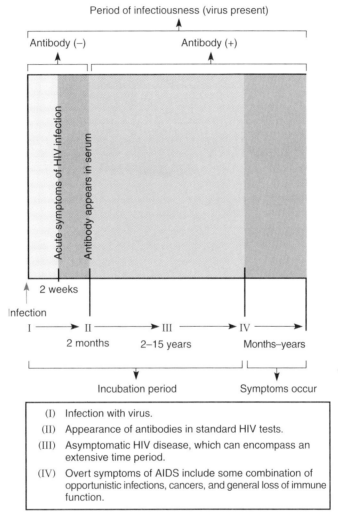

Period of infectiousness (virus present)

Antibody (−)   Antibody (+)

Acute symptoms of HIV infection

Antibody appears in serum

2 weeks

Infection

I ——→ II ——→ III ——→ IV ——→

2 months   2–15 years   Months–years

Incubation period   Symptoms occur

(I)	Infection with virus.
(II)	Appearance of antibodies in standard HIV tests.
(III)	Asymptomatic HIV disease, which can encompass an extensive time period.
(IV)	Overt symptoms of AIDS include some combination of opportunistic infections, cancers, and general loss of immune function.

Stages in HIV infection, AIDS
Figure 25.16

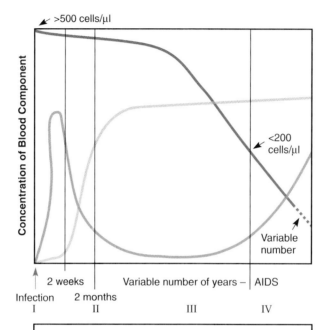

Level of virus antigen
Level of antibodies to one or more antigens
Level of CD4 T cells

>500 cells/µl

Concentration of Blood Component

<200 cells/µl

Variable number

2 weeks   Variable number of years —   AIDS

Infection
        2 months
I   II   III   IV

A comparison of blood levels of viruses, antibodies, and T cells covering the same time frame depicted in figure 25.16. Virus levels are high during the initial acute infection and decrease until the later phases of HIV disease and AIDS. Antibody levels gradually rise and remain relatively high throughout phases III and IV. T-cell numbers remain relatively normal until the later phases of HIV disease and full-blown AIDS.

Dynamics of virus antigen, antibody, and T cells in circulation
Figure 25.17

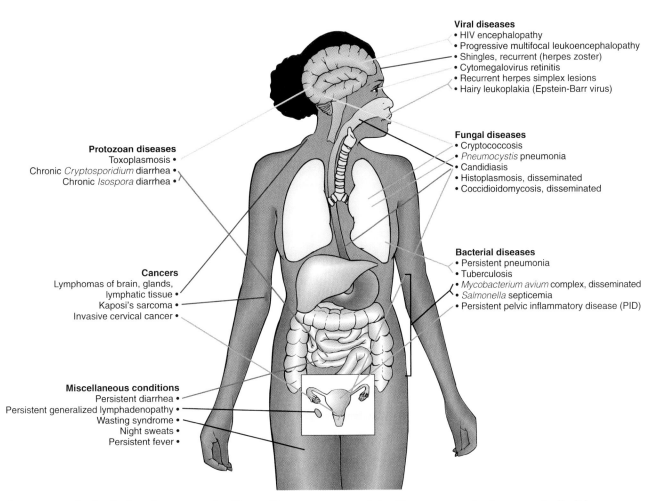

Viral diseases
• HIV encephalopathy
• Progressive multifocal leukoencephalopathy
• Shingles, recurrent (herpes zoster)
• Cytomegalovirus retinitis
• Recurrent herpes simplex lesions
• Hairy leukoplakia (Epstein-Barr virus)

Protozoan diseases
Toxoplasmosis •
Chronic *Cryptosporidium* diarrhea •
Chronic *Isospora* diarrhea •

Fungal diseases
• Cryptococcosis
• *Pneumocystis* pneumonia
• Candidiasis
• Histoplasmosis, disseminated
• Coccidioidomycosis, disseminated

Cancers
Lymphomas of brain, glands,
lymphatic tissue •
Kaposi's sarcoma •
Invasive cervical cancer •

Bacterial diseases
• Persistent pneumonia
• Tuberculosis
• *Mycobacterium avium* complex, disseminated
• *Salmonella* septicemia
• Persistent pelvic inflammatory disease (PID)

Miscellaneous conditions
Persistent diarrhea •
Persistent generalized lymphadenopathy •
Wasting syndrome •
Night sweats •
Persistent fever •

Opportunistic infections and other diseases used in the expanded case definition of AIDS
Figure 25.18

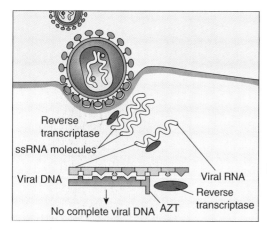

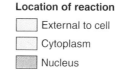

Location of reaction

☐ External to cell

☐ Cytoplasm

▨ Nucleus

(a) A prominent group of drugs (AZT, ddl, 3TC) are nucleoside analogs that inhibit reverse transcriptase. They are inserted in place of the natural nucleotide by reverse transcriptase but block further action of the enzyme and synthesis of viral DNA.

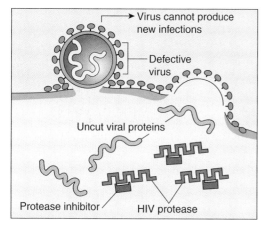

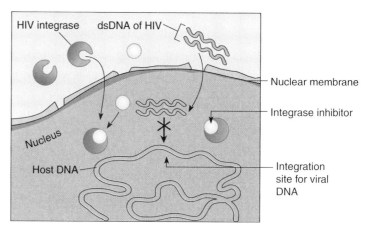

(b) Protease inhibitors plug into the active sites on HIV protease. This enzyme is necessary to cut elongate HIV protein strands and produce functioning smaller protein units. Because the enzyme is blocked, the proteins remain uncut, and abnormal defective viruses are formed.

(c) Integrase inhibitors are a new class of experimental drugs that attach to the enzyme required to splice the dsDNA from HIV into the host genome. This will prevent formation of the provirus and block future virus multiplication in that cell.

Mechanisms of action of anti-HIV drugs
Figure 25.20

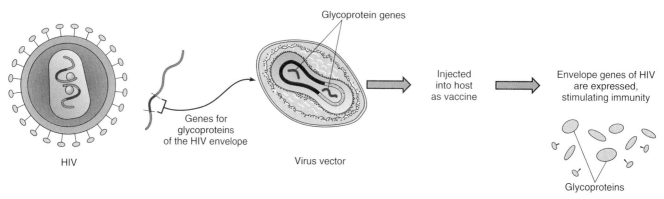

"Trojan horse," or viral vector technique, a novel technique for making an AIDS vaccine
Figure 25.21

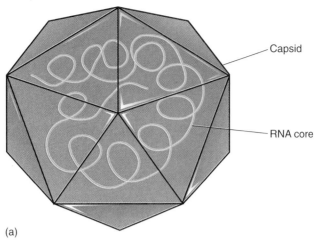

(a)

Typical structure of a picornavirus
Figure 25.23

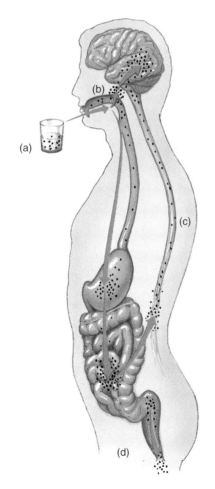

(b)

(a)

(c)

(d)

**The stages of infection
and pathogenesis of
poliomyelitis**
Figure 25.24

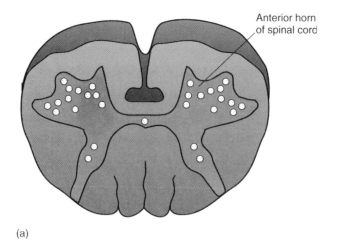

(a)

Targets of poliovirus
Figure 25.25

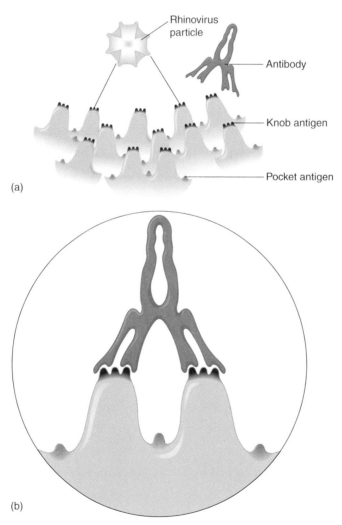

Rhinovirus particle

Antibody

Knob antigen

Pocket antigen

(a)

(b)

Structure of a rhinovirus
Figure 25.27

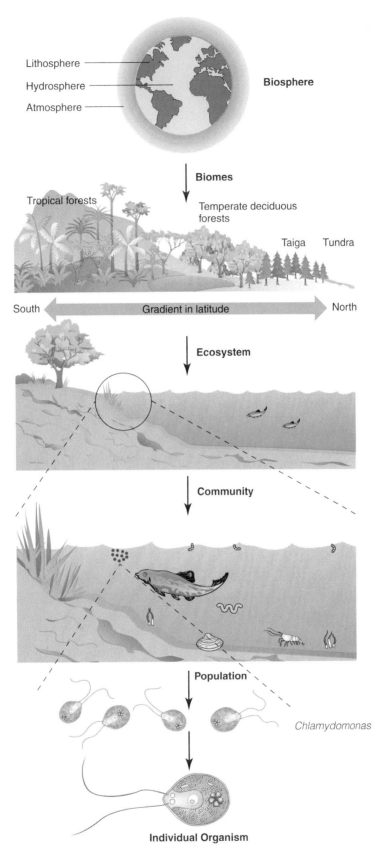

Levels of organization in an ecosystem, ranging from the biosphere to the individual organism

Figure 26.1

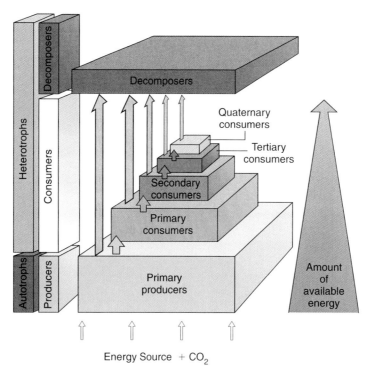

A trophic and energy pyramid
Figure 26.2

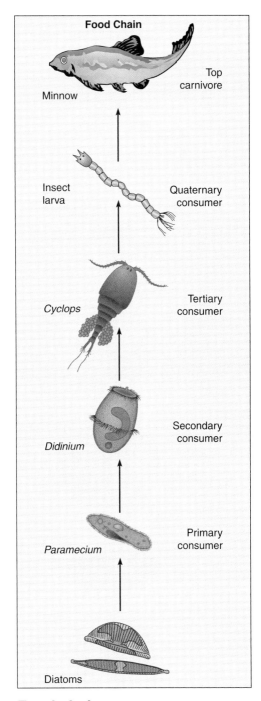

Food chain
Figure 26.3

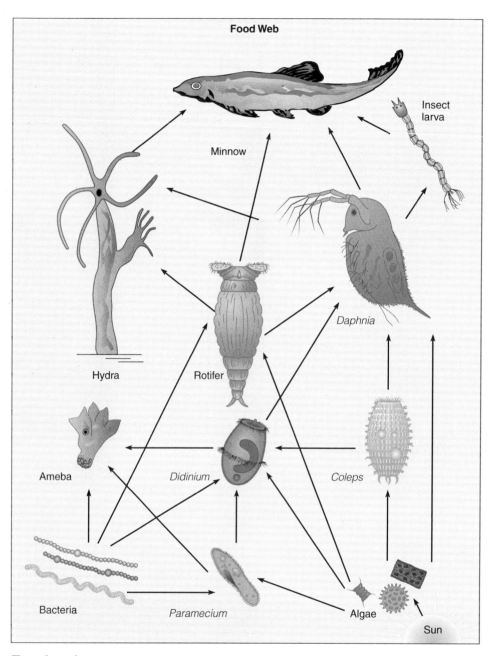

Food web

Figure 26.4

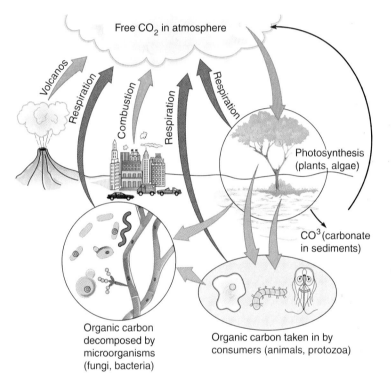

Free CO$_2$ in atmosphere

Volcanos

Respiration

Combustion

Respiration

Respiration

Photosynthesis
(plants, algae)

CO^3 (carbonate
in sediments)

Organic carbon
decomposed by
microorganisms
(fungi, bacteria)

Organic carbon taken in by
consumers (animals, protozoa)

The carbon cycle
Figure 26.5

Summary equation:

$$6CO_2 + 6H_2O \xrightarrow{\text{Light and pigment}} \underset{\text{Glucose}}{C_6H_{12}O_6} + 6O_2 \uparrow$$

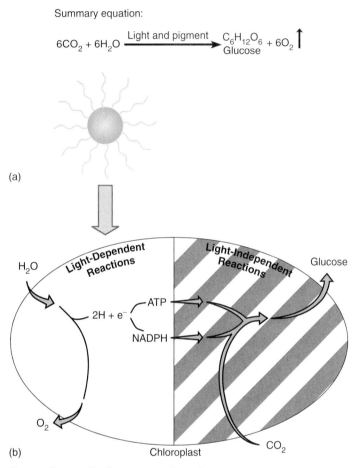

(a)

(b)

Overview of photosynthesis
Figure 26.6

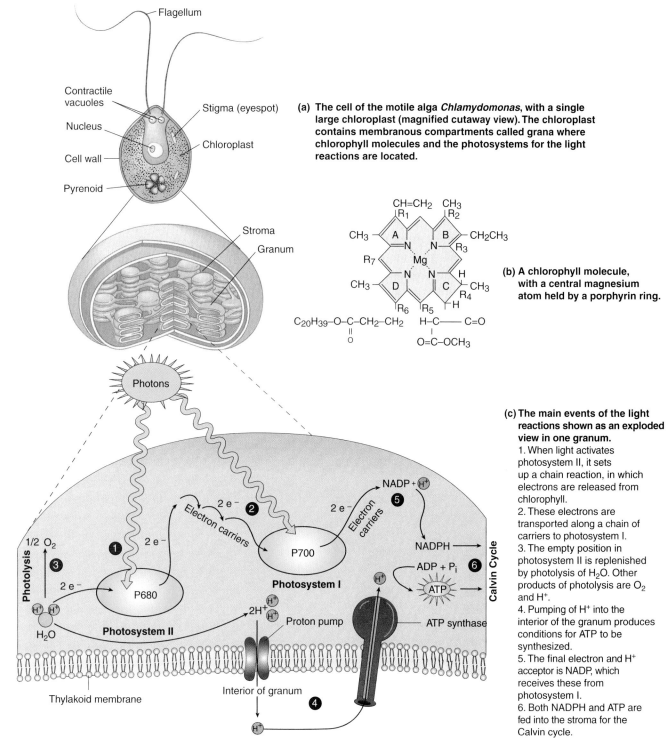

(a) The cell of the motile alga *Chlamydomonas*, with a single large chloroplast (magnified cutaway view). The chloroplast contains membranous compartments called grana where chlorophyll molecules and the photosystems for the light reactions are located.

(b) A chlorophyll molecule, with a central magnesium atom held by a porphyrin ring.

(c) The main events of the light reactions shown as an exploded view in one granum.
1. When light activates photosystem II, it sets up a chain reaction, in which electrons are released from chlorophyll.
2. These electrons are transported along a chain of carriers to photosystem I.
3. The empty position in photosystem II is replenished by photolysis of H_2O. Other products of photolysis are O_2 and H^+.
4. Pumping of H^+ into the interior of the granum produces conditions for ATP to be synthesized.
5. The final electron and H^+ acceptor is NADP, which receives these from photosystem I.
6. Both NADPH and ATP are fed into the stroma for the Calvin cycle.

The reactions of photosynthesis
Figure 26.7

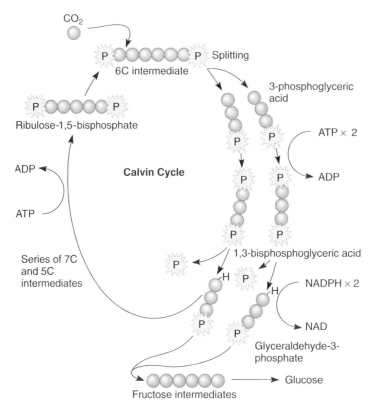

The Calvin cycle
Figure 26.8

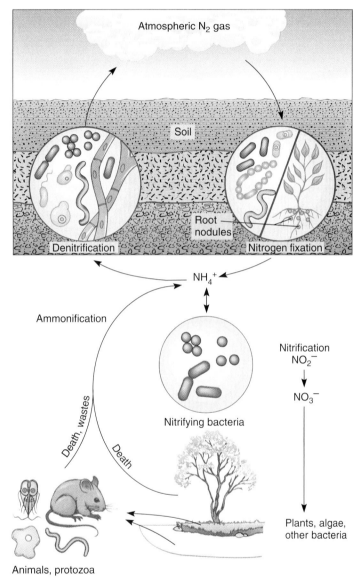

Atmospheric N_2 gas

Soil

Root
nodules

Denitrification

Nitrogen fixation

NH_4^+

Ammonification

Nitrification
NO_2^-

NO_3^-

Nitrifying bacteria

Death, wastes

Death

Plants, algae,
other bacteria

Animals, protozoa

The simplified events in the nitrogen cycle
Figure 26.9

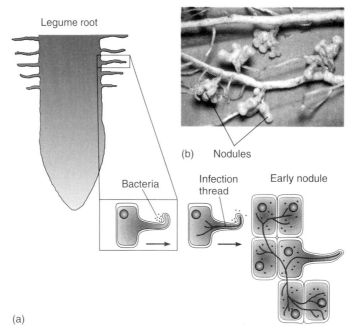

Legume root

(b) Nodules

Bacteria

Infection thread

Early nodule

(a)

Nitrogen fixation through symbiosis
Figure 26.10

b: © John D. Cunningham/Visuals Unlimited

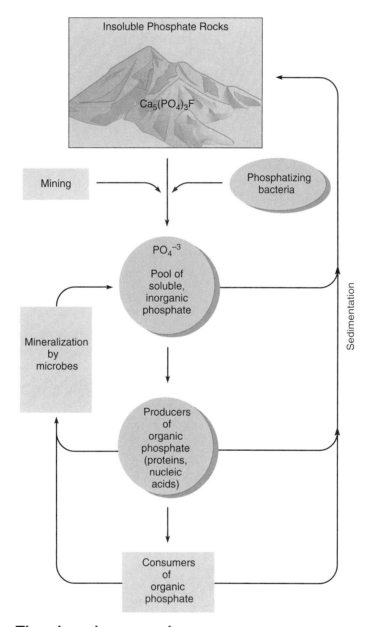

The phosphorus cycle
Figure 26.12

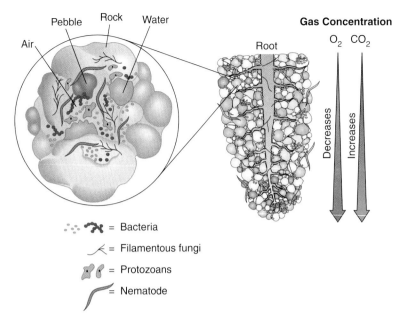

= Bacteria

= Filamentous fungi

= Protozoans

= Nematode

The structure of the rhizosphere and the microhabitats that develop in response to soil particles, moisture, air, and gas content
Figure 26.14

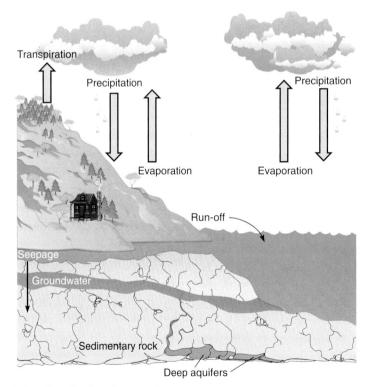

The hydrologic cycle
Figure 26.16

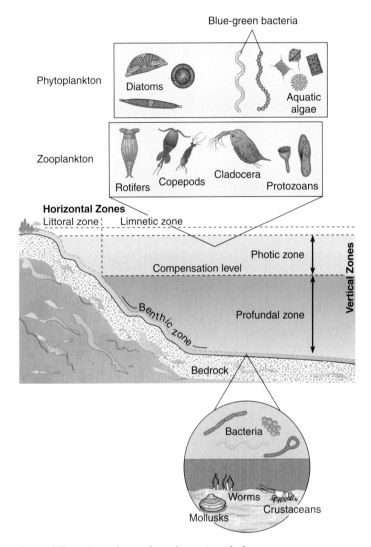

Stratification in a freshwater lake
Figure 26.17

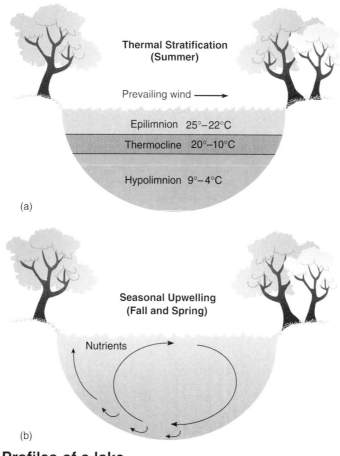

Thermal Stratification
(Summer)

Prevailing wind ⟶

Epilimnion 25°–22°C

Thermocline 20°–10°C

Hypolimnion 9°–4°C

(a)

Seasonal Upwelling
(Fall and Spring)

Nutrients

(b)

Profiles of a lake
Figure 26.18

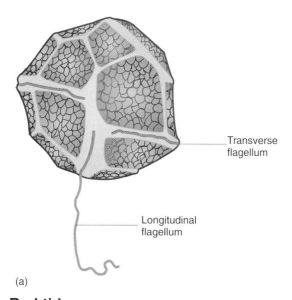

Transverse
flagellum

Longitudinal
flagellum

(a)

Red tides
Figure 26.19

(a, b) Membrane filter technique. (a) The water sample is filtered through a sterile membrane filter assembly and collected in a flask.

(b) The filter is removed and placed in a small Petri dish containing a differential selective medium such as M-FD endo broth and incubated.

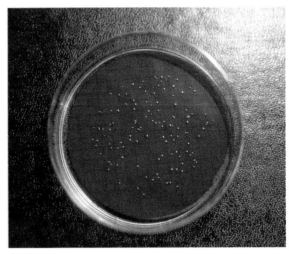

(c) On M-FD endo medium, colonies of *Escherichia coli* often yield a noticeable metallic sheen. The medium permits easy differentiation of various genera of coliforms, and the grid pattern can be used as a guide for rapidly counting the colonies.

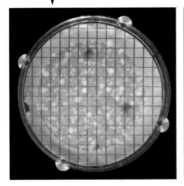

Total coliforms fluoresce under a black light.

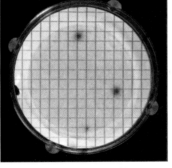

E. coli colonies are blue under natural light.

(d) Some tests for water-borne coliforms are based on formation of specialized enzymes to metabolize lactose. The MI tests shown here utilize synthetic substrates that release a colored substance when the appropriate enzymes are present. The total coliform count is indicated by the plate on the left; fecal coliforms (*E. coli*) are seen in the plate on the right. This test is especially accurate on surface or groundwater samples.

Rapid methods of water analysis for coliform contamination
Figure 26.21

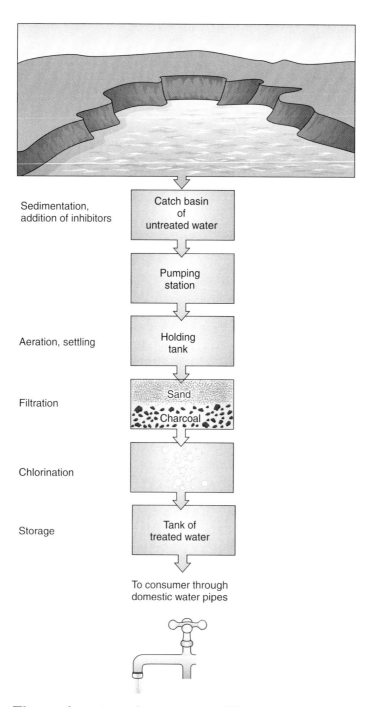

Sedimentation,
addition of inhibitors

Catch basin
of
untreated water

Pumping
station

Aeration, settling

Holding
tank

Filtration

Sand

Charcoal

Chlorination

Storage

Tank of
treated water

To consumer through
domestic water pipes

The major steps in water purification as carried out by a modern municipal treatment plant
Figure 26.22

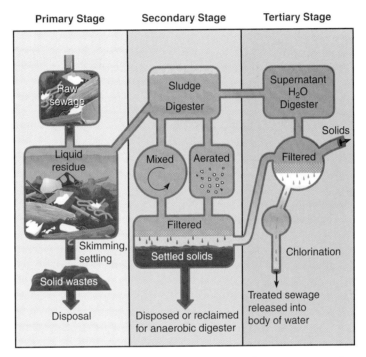

Primary Stage **Secondary Stage** **Tertiary Stage**

Raw sewage

Liquid residue

Skimming, settling

Solid wastes

Disposal

Sludge Digester

Mixed

Aerated

Filtered

Settled solids

Disposed or reclaimed for anaerobic digester

Supernatant H₂O Digester

Solids

Filtered

Chlorination

Treated sewage released into body of water

The primary, secondary, and tertiary stages in sewage treatment
Figure 26.23

Processing Step		Biological Change
Barley moistening and germination		Enzymatic release of soluble carbohydrates
	Malting floor	
Drying and crushing		
Mashing	Mash tun	Further enzymatic activity—release of maltose, dextrins, and proteins
Add hops		
Heat in brew kettle	Brew kettle	Enzyme inactivation Flavoring from hops Clarification
Add yeast		Remove hops
Fermentation		Alcoholic fermentation
Storage (lagering)		Final flavor development
Packaging		

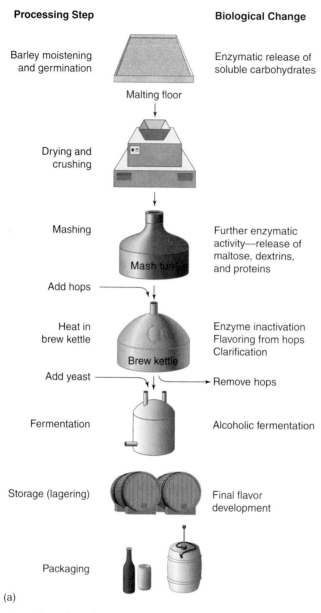

(a)

Brewing basics
Figure 26.25

Processing Step		Outcome
Grape pressing		Formation of must with fruit sugars
Heat sterilization		
Yeast inoculation		Elimination of contaminants Addition of desired organisms
Fermentation of must	Tank	Alcohol production from sugars
Storage in barrels to age	Barrel	Development of final wine bouquet
Filtration and collection		Removal of yeast and particles
Bottling		

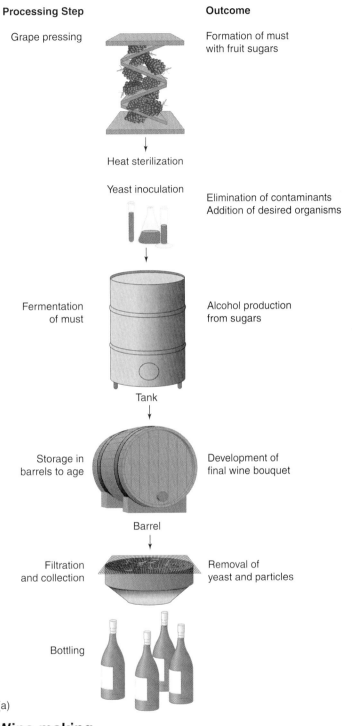

(a)

Wine making
Figure 26.27

Processing Step

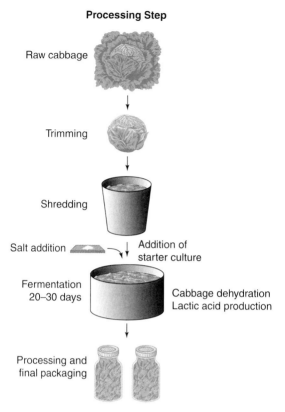

Raw cabbage

Trimming

Shredding

Salt addition

Addition of
starter culture

Fermentation
20–30 days

Cabbage dehydration
Lactic acid production

Processing and
final packaging

**Steps in the preparation of
sauerkraut**
Figure 26.28

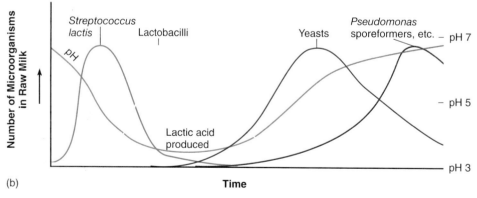

(b)

Microbes at work in milk products
Figure 26.29

Chart from Philip L. Carpenter, *Microbiology,* 3rd ed., copyright © 1972 by Holt, Rinehart
and Winston, Inc. Reprinted by permission of the publisher.

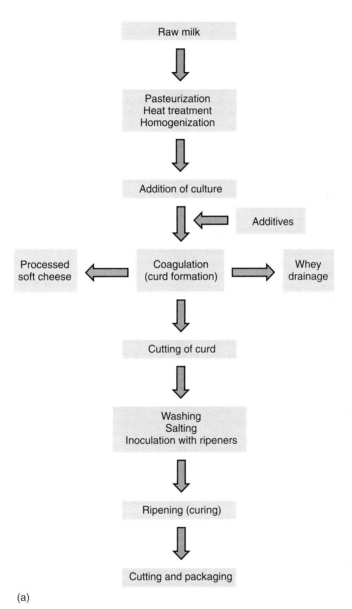

(a)

Cheese making
Figure 26.30

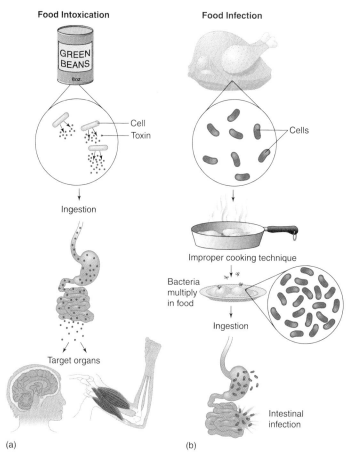

Food Intoxication

GREEN
BEANS
8oz

Cell
Toxin

Ingestion

Target organs

(a)

Food Infection

Cells

Improper cooking technique

Bacteria
multiply
in food

Ingestion

Intestinal
infection

(b)

Food-borne illnesses of microbial origin
Figure 26.31

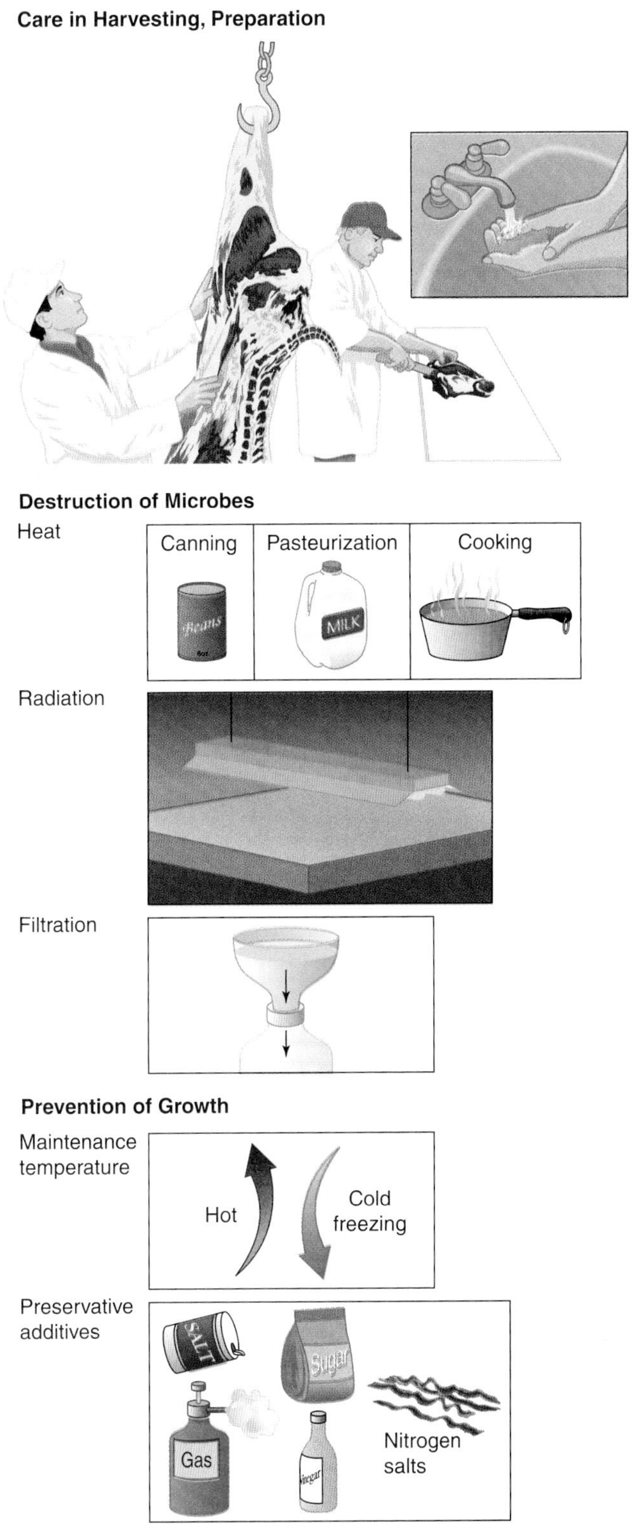
Care in Harvesting, Preparation
Destruction of Microbes
Heat
Canning
Beans
Pasteurization
MILK
Cooking
Radiation
Filtration
Prevention of Growth
Maintenance temperature
Hot
Cold freezing
Preservative additives
SALT
Sugar
Gas
Nitrogen salts

The primary methods to prevent food poisoning and food spoilage
Figure 26.32

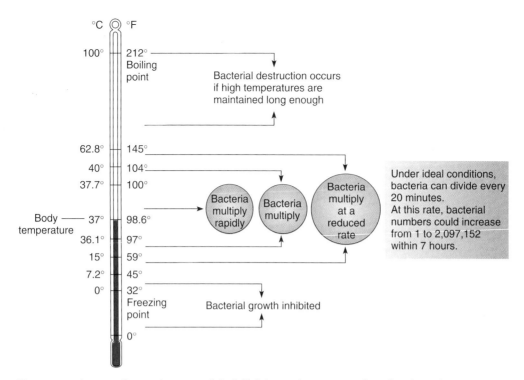

Temperatures favoring and inhibiting the growth of microbes in food
Figure 26.34

From Ronald Atlas, *Microbiology: Fundamentals and Applications,* 2nd ed., © 1998, p. 475.
Reprinted by permission of Prentice Hall, Upper Saddle River, New Jersey.

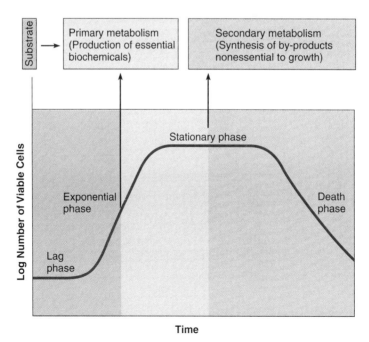

The origins of primary and secondary microbial metabolites harvested by industrial processes
Figure 26.35

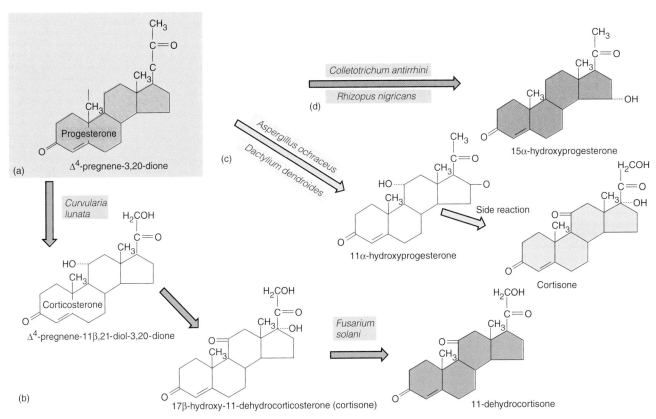

(a) Progesterone
Δ⁴-pregnene-3,20-dione

(b) Curvularia lunata
Corticosterone
Δ⁴-pregnene-11β,21-diol-3,20-dione
17β-hydroxy-11-dehydrocorticosterone (cortisone)

Fusarium solani
11-dehydrocortisone

(c) Aspergillus ochraceus
Dactylium dendroides
11α-hydroxyprogesterone
Side reaction
Cortisone

(d) Colletotrichum antirrhini
Rhizopus nigricans
15α-hydroxyprogesterone

An example of biotransformation by microorganisms in the industrial production of steroid hormones

Figure 26.36

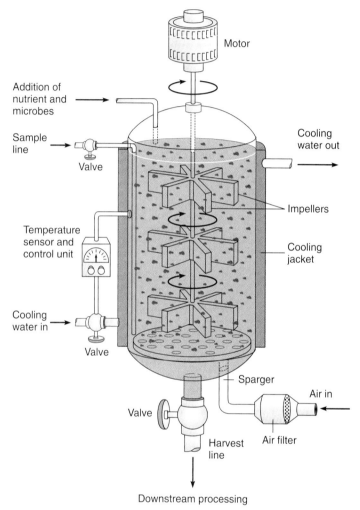

Motor

Addition of
nutrient and
microbes

Sample
line

Valve

Cooling
water out

Impellers

Temperature
sensor and
control unit

Cooling
jacket

Cooling
water in

Valve

Sparger

Air in

Valve

Air filter

Harvest
line

Downstream processing

**A schematic diagram of an industrial
fermentor for mass culture of microorganisms**
Figure 26.38

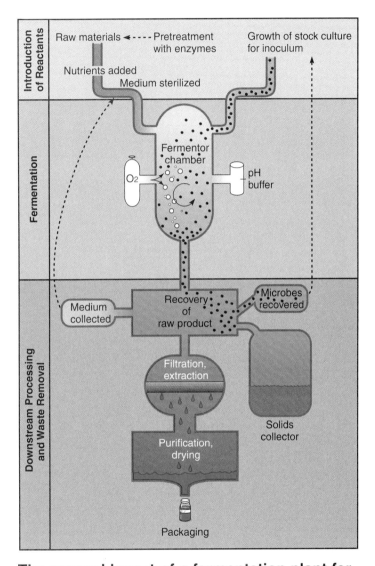

The general layout of a fermentation plant for industrial production of drugs, enzymes, fuels, vitamins, and amino acids
Figure 26.39

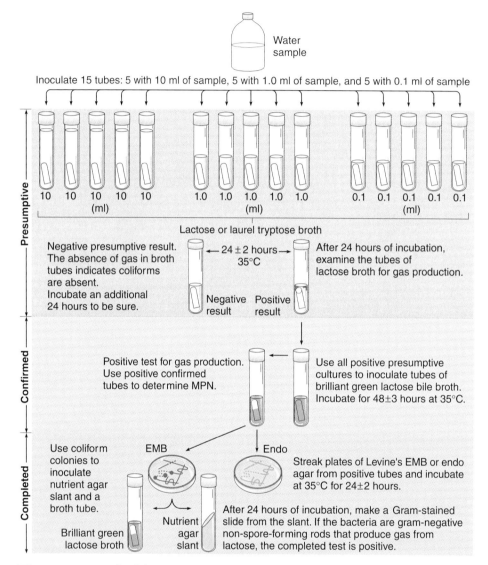

Water sample

Inoculate 15 tubes: 5 with 10 ml of sample, 5 with 1.0 ml of sample, and 5 with 0.1 ml of sample

Presumptive

10	10	10	10	10
(ml)

1.0	1.0	1.0	1.0	1.0
(ml)

0.1	0.1	0.1	0.1	0.1
(ml)

Lactose or laurel tryptose broth

Negative presumptive result. The absence of gas in broth tubes indicates coliforms are absent. Incubate an additional 24 hours to be sure.

← 24 ± 2 hours → 35°C

Negative result   Positive result

After 24 hours of incubation, examine the tubes of lactose broth for gas production.

Confirmed

Positive test for gas production. Use positive confirmed tubes to determine MPN.

Use all positive presumptive cultures to inoculate tubes of brilliant green lactose bile broth. Incubate for 48±3 hours at 35°C.

Completed

Use coliform colonies to inoculate nutrient agar slant and a broth tube.

EMB

Endo

Streak plates of Levine's EMB or endo agar from positive tubes and incubate at 35°C for 24±2 hours.

Brilliant green lactose broth

Nutrient agar slant

After 24 hours of incubation, make a Gram-stained slide from the slant. If the bacteria are gram-negative non-spore-forming rods that produce gas from lactose, the completed test is positive.

The most probable number (MPN) procedure for determining the coliform content of a water sample
Figure A.2

Notes

Notes

Notes

Notes

Notes

Notes

Notes

Notes

Notes

Notes

Notes